Dorothy Warren

REVISION PLUS

OCR Twenty First Century GCSE Chemistry A

Re[...]om Companion

Ideas about Science

Introduction to Ideas about Science

The OCR Twenty First Century Chemistry specification aims to ensure that you develop an understanding of science itself – of how scientific knowledge is obtained, the kinds of evidence and reasoning behind it, its strengths and limitations, and how far we can rely on it.

These issues are explored through Ideas about Science, which are built into the specification content and summarised over the following pages.

The tables below give an overview of the Ideas about Science that can be assessed in each unit and provide examples of content which support them in this guide.

Unit A171 (Modules C1, C2 and C3)

Ideas about Science	Example of Supporting Content
Data: their importance and limitations	Data about Pollution (pages 4–5)
Cause–effect explanations	Identifying Health Hazards (page 8)
Developing scientific explanations	The Origins of Mineral Wealth in Britain (page 17)
The scientific community	Identifying Health Hazards (page 8)
Risk	Food and the Government (page 22); Safe and Sustainable Chemicals (page 25)
Making decisions about science and technology	Evaluating Nanomaterials (page 16); Life Cycle Assessment (LCA) (page 26)

Unit A172 (Modules C4, C5 and C6)

Ideas about Science	Example of Supporting Content
Data: their importance and limitations	Titration; Interpreting Results (pages 58–59)
Cause–effect explanations	Rates of Reactions; Measuring the Rate of Reaction; Analysing the Rate of Reaction; Changing the Rate of Reaction (pages 59–62)
Developing scientific explanations	The Development of the Periodic Table (page 29); Spectroscopy (page 31)
The scientific community	The Development of the Periodic Table (page 29)
Risk	Hazard Symbols; Safety Precautions; Group 1 – The Alkali Metals; Group 7 – The Halogens (pages 33–36)
Making decisions about science and technology	Metals and the Environment (page 48)

Unit A173 (Module C7)

Ideas about Science	Example of Supporting Content
Data: their importance and limitations	Quantitative Analysis by Titration; Interpreting Titration Results; Evaluating Experimental Results (pages 85–87)
Cause–effect explanations	Quantitative Analysis by Titration; Interpreting Titration Results; Evaluating Experimental Results (pages 85–87)
Developing scientific explanations	Looking to the Future (page 79)
The scientific community	Looking to the Future (page 79)
Risk	Green Chemistry (pages 66–67)
Making decisions about science and technology	Green Chemistry (pages 66–67); Ammonia is a Very Important Chemical (page 77)

1 Data: Their Importance and Limitations

Science is built on data. Chemists carry out experiments to collect and interpret data, seeing whether the data agree with their explanations. If the data do agree, then it means the current explanation is more likely to be correct. If not, then the explanation has to be changed.

Experiments aim to find out what the 'true' value of a quantity is. Quantities are affected by errors made when carrying out the experiment and random variation. This means that the measured value may be different to the true value. Chemists try to control all the factors that could cause this uncertainty.

Chemists always take repeat readings to try to make sure that they have accurately estimated the true value of a quantity. The mean is calculated and is the best estimate of what the true value of a quantity is. The more times an experiment is repeated, the greater the chance that a result near to the true value will fall within the mean.

The range, or spread, of data gives an indication of where the true value must lie. Sometimes a measurement will not be in the zone where the majority of readings fall. It may look like the result (called an 'outlier') is wrong – however, it does not automatically mean that it is. The outlier has to be checked by repeating the measurement of that quantity. If the result cannot be checked, then it should still be used.

Here is an example of an outlier in a set of data:

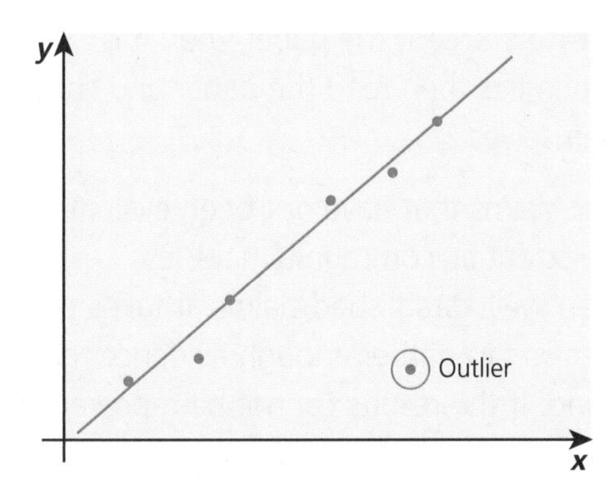

The spread of the data around the mean (the range) gives an idea of whether it really is different to the mean from another measurement. If the ranges for each mean do not overlap, then it is more likely that the two means are different. However, sometimes the ranges do overlap and there may be no significant difference between them.

The ranges also give an indication of reliability – a wide range makes it more difficult to say with certainty that the true value of a quantity has been measured. A small range suggests that the mean is closer to the true value.

If an outlier is discovered, you need to be able to defend your decision as to whether you keep it or discard it.

2 Cause–effect Explanations

Science is based on the idea that a factor has an effect on an outcome. Chemists make predictions as to how the input variable will change the outcome variable. To make sure that only the input variable can affect the outcome, chemists try to control all the other variables that could potentially alter it. This is called 'fair testing'.

You need to be able to explain why it is necessary to control all the factors that might affect the outcome. This means suggesting how they could influence the outcome of the experiment.

A correlation is where there is an apparent link between a factor and an outcome. It may be that as the factor increases, the outcome increases as well. On the other hand, it may be that when the factor increases, the outcome decreases.

For example, there is a correlation between temperature and the rate of rusting – the higher the temperature, the more rapid the rate of rusting.

Just because there is a correlation does not necessarily mean that the factor causes the outcome. Further experiments are needed to establish this. It could be that another factor causes the outcome or that both the original factor and outcome are caused by something else.

The following graph suggests a correlation between going to the opera regularly and living longer. It is far more likely that if you have the money to go to the opera, you can afford a better diet and health care. Going to the opera is not the true cause of the correlation.

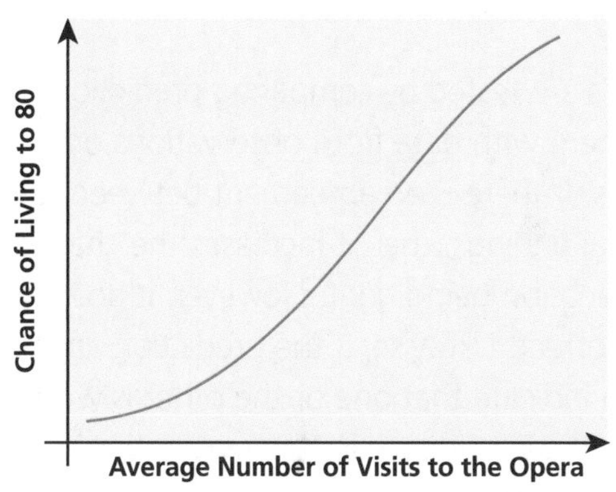

Sometimes the factor may alter the chance of an outcome occurring but does not guarantee it will lead to it. The statement 'the more time spent on a sun bed the greater the chance of developing skin cancer' is an example of this type of correlation, as some people will not develop skin cancer even if they do spend a lot of time on a sun bed.

To investigate claims that a factor increases the chance of an outcome, chemists have to study groups of people who either share as many factors as possible or are chosen randomly to try to ensure that all factors will present in people in the test group. The larger the experimental group, the more confident chemists can be about the conclusions made.

> HT Even so, a correlation and cause will still not be accepted by chemists unless there is a scientific mechanism that can explain them.

3 Developing Scientific Explanations

Chemists devise hypotheses (predictions of what will happen in an experiment), along with an explanation (the scientific mechanism behind the hypotheses) and theories (that can be tested).

Explanations involve thinking creatively to work out why data have a particular pattern. Good scientific explanations account for most or all of the data already known. Sometimes they may explain a range of phenomena that were not previously thought to be linked. Explanations should enable predictions to be made about new situations or examples.

When deciding on which is the better of two explanations, you should be able to give reasons why.

Explanations are tested by comparing predictions based on them with data from observations or experiments. If there is an agreement between the experimental findings, then it increases the chance of the explanation being right. However, it does not prove it is correct. Likewise, if the prediction and observation indicate that one or the other is wrong, it decreases the confidence in the explanation on which the prediction is based.

4 The Scientific Community

Once a chemist has carried out enough experiments to back up his/her claims, they have to be reported. This enables the scientific community to carefully check the claims, something which is required before they are accepted as scientific knowledge.

Chemists attend conferences where they share their findings and sound out new ideas and explanations. This can lead to chemists revisiting their work or developing links with other laboratories to improve it.

The next step is writing a formal scientific paper and submitting it to a journal in the relevant field. The paper is allocated to peer reviewers (experts in their field), who carefully check and evaluate the paper. If the peer reviewers accept the paper, then it is published. Chemists then read the paper and check the work themselves.

New scientific claims that have not been evaluated by the whole scientific community have less credibility than well-established claims. It takes time for other chemists to gather enough evidence that a theory is sound. If the results cannot be repeated or replicated by themselves or others, then chemists will be sceptical about the new claims.

If the explanations cannot be arrived at from the available data, then it is fair and reasonable for different chemists to come up with alternative explanations. These will be based on the background and experience of the chemists. It is through further experimentation that the best explanation will be chosen.

This means that the current explanation has the greatest support. New data are not enough to topple it. Only when the new data are sufficiently repeated and checked will the original explanation be changed.

> HT You need to be able to suggest reasons why an accepted explanation will not be given up immediately when new data, which appear to conflict with it, have been published.

5 Risk

Everything we do (or not do) carries risk. Nothing is completely risk-free. New technologies and processes based on scientific advances often introduce new risks.

Risk is sometimes calculated by measuring the chance of something occurring in a large sample over a given period of time (calculated risk).
This enables people to take informed decisions about whether the risk is worth taking. In order to

decide, you have to balance the benefit (to individuals or groups) with the consequences of what could happen.

For example, deciding whether or not to add chlorine to drinking water involves weighing up the benefit (of reducing the spread of cholera) against the risk of a toxic chlorine leak at the purification plant.

Risk which is associated with something that someone has chosen to do is easier to accept than risk which has been imposed on them.

HT Perception of risk changes depending on our personal experience (perceived risk). Familiar risks (e.g. using bleach without wearing gloves) tend to be under-estimated, whilst unfamiliar risks (e.g. making chlorine in the laboratory) and invisible or long-term risks (e.g. cleaning up mercury from a broken thermometer) tend to be over-estimated.

For example, many people under-estimate the risk that adding limescale remover and bleach to a toilet at the same time might produce toxic chlorine gas.

Governments and public bodies try to assess risk and create policy on what is and what is not acceptable. This can be controversial, especially when the people who benefit most are not the ones at risk.

6 Making Decisions about Science and Technology

Science has helped to create new technologies that have improved the world, benefiting millions of people. However, there can be unintended consequences of new technologies, even many decades after they were first introduced. These could be related to the impact on the environment or to the quality of life.

When introducing new technologies, the potential benefits must be weighed up against the risks.

Sometimes unintended consequences affecting the environment can be identified. By applying the scientific method (making hypotheses, explanations and carrying out experiments), chemists can devise new ways of putting right the impact. Devising life cycle assessments helps chemists to try to minimise unintended consequences and ensure sustainability.

Some areas of chemistry could have a high potential risk to individuals or groups if they go wrong or if they are abused. In these areas the Government ensures that regulations are in place.

The scientific approach covers anything where data can be collected and used to test a hypothesis. It cannot be used when evidence cannot be collected (e.g. it cannot test beliefs or values).

Just because something can be done does not mean that it should be done. Some areas of scientific research or the technologies resulting from them have ethical issues associated with them. This means that not all people will necessarily agree with it.

Ethical decisions have to be made, taking into account the views of everyone involved, whilst balancing the benefits and risks.

It is impossible to please everybody, so decisions are often made on the basis of which outcome will benefit most people. Within a culture there will also be some actions that are always right or wrong, no matter what the circumstances are.

Contents

Module C1 (Air Quality)

Air pollutants can affect the environment and our health. However, there are options available for improving air quality in the future. This module looks at:

- the chemicals that make up air and the ones that are pollutants
- data about air pollution
- the chemical reactions that produce air pollutants
- what happens to pollutants in the atmosphere
- the steps that can be taken to improve air quality.

Chemicals in the Air

The Earth is surrounded by a thin layer of gases called the **atmosphere**. Air forms part of the atmosphere. It is a mixture of different gases consisting of small molecules with large spaces between them. Air contains about 78% **nitrogen**, 21% **oxygen**, 1% **argon and other noble gases**. There are also small amounts of **water vapour**, **carbon dioxide** and **particulates**. The amount of water vapour and polluting gases varies as a result of human activity or by natural processes (e.g. volcanoes).

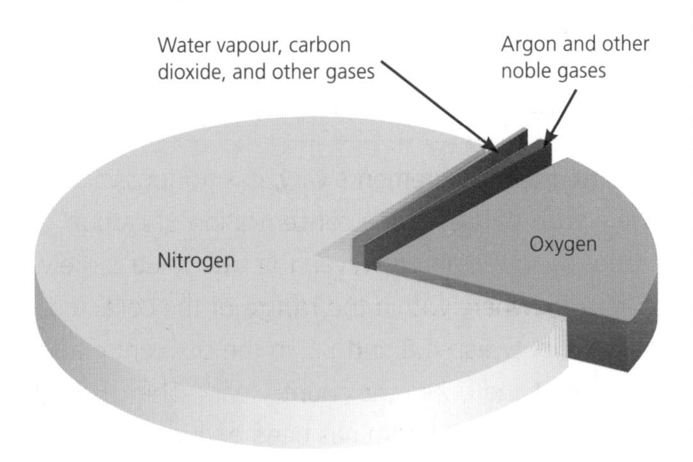

The Earth's Atmosphere

Since the formation of the Earth, 4.6 billio ago, the atmosphere has changed a lot. Th timescale, however, is enormous because o years is one thousand million (1 000 000 00

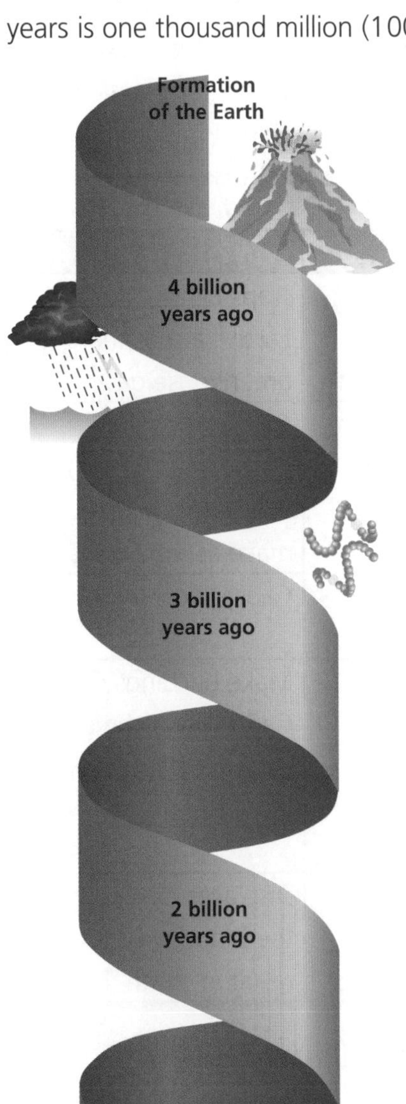

The earliest atmosp contained ammonia water vapour and ca dioxide. These gases came from inside the Earth and were often released through the action of volcanoes.

As the temperature of the planet fell, the water vapour in the atmosphere condensed to form the oceans and seas.

The evolution of photosynthesising organisms started to reduce the amount of carbon dioxide and increase the amount of oxygen in the atmosphere.

Carbon from carbon dioxide in the air became locked up in sedimentary rocks as carbonates and fossil fuels after dissolving in the oceans.

Clean air contains about:

- **78% nitrogen**
- **21% oxygen**
- **1% other gases, including 0.035% carbon dioxide**.

Normal air contains varying amounts of **water vapour** and some **polluting gases**. The variation in the quantities of these gases is partly due to human activities.

Air Quality

ts in the Air

, are chemicals that can harm the
nt and our health.

, that harm the environment can also harm
indirectly. For example, acid rain makes the
n rivers and lakes too acidic for plants and
nals to survive. This has a direct impact on our
d chain and natural resources like trees.

Pollutant	Harmful to...	Why?
Carbon dioxide	Environment	Traps heat in the Earth's atmosphere (a greenhouse gas).
Nitrogen oxides	Environment Humans	Cause acid rain. Cause breathing problems and can make asthma worse.
Sulfur dioxide	Environment	Causes acid rain.
Particulates (small particles of solids, e.g. carbon)	Environment Humans	Make buildings dirty. Can make asthma and lung infections worse if inhaled.
Carbon monoxide	Humans	Displaces oxygen in the blood, which can result in death.

Measuring Pollutants

By measuring the **concentrations** of pollutants in the air, it is possible to assess air quality. The units of measurement used are **ppb** (**parts per billion**) or **ppm** (**parts per million**). For example, a sulfur dioxide concentration of 16ppb means that in every one billion (1 000 000 000) **molecules** of air, 16 will be sulfur dioxide molecules.

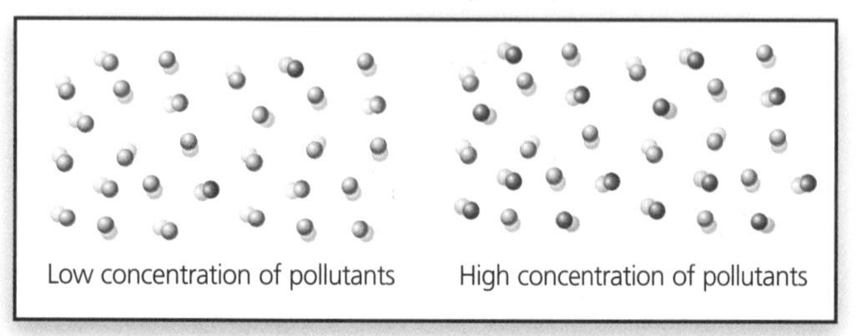

Data about Pollution

Data is very important to scientists because it can be used to test a theory or explanation.

Example
One theory states that carbon monoxide (CO) is an example of a pollutant caused by human activity.

If this is true, carbon monoxide concentrations are likely to be higher in densely populated areas, e.g. cities.

The data below was collected on the same day using a carbon monoxide meter:

Location	Time	Carbon Monoxide Concentration (ppm)
City centre	9.00am	5.2
	10.00am	4.9
	11.00am	5.0
	12.00pm	2.6
	1.00pm	4.8
Country park	9.00am	0.2
	10.00am	0.1
	11.00am	0.1
	12.00pm	0.0
	1.00pm	0.1

Measurements like this can vary because:

- **variables** (factors that change), like the volume of traffic and weather, affect concentrations
- all measuring **equipment** has limited accuracy
- the user's **skill** will affect the accuracy of the measurement.

Because the measurements vary, it is not possible to give a **true value** for the concentration of carbon monoxide in the air. However, the true value is likely to lie somewhere within the **range** of the collected data, i.e. between 4.8 and 5.2 in the city centre and between 0 and 0.2 in the country park. The measurement of 2.6ppm has been excluded from the data range for the city centre because it is an **outlier**. Outliers are measurements that stand out as being very different from the rest of the data. They normally indicate some sort of error.

HT You must be able to say why 2.6ppm is an outlier, e.g. the operator may have misread the scale. It is unlikely that the volume of traffic would have decreased at midday. In fact, you might expect it to increase as people leave their workplaces for lunch.

It is important that measurements are repeated. If you look at one measurement on its own, you cannot tell if it is reliable. However, if you look at lots of repeated measurements, any errors should stand out.

By calculating the **mean** (finding the average) of a set of repeated measurements, you can overcome small variations and get a **best estimate** of the true value.

$$\textbf{Mean} = \frac{\textbf{Sum of all values}}{\textbf{Number of values}}$$

Do not use outliers in mean calculations!

$$\text{City} = \frac{5.2 + 4.9 + 5.0 + 4.8}{4} = \textbf{5.0ppm}$$

$$\text{Country} = \frac{0.2 + 0.1 + 0.1 + 0.0 + 0.1}{5} = \textbf{0.1ppm}$$

The mean carbon monoxide concentration in the city centre is significantly higher than the mean carbon monoxide concentration in the country park. So, this data supports the theory that carbon monoxide is a pollutant caused by human activity.

In fact, about half of all carbon monoxide emissions in the UK are produced by road transport, with the rest coming from homes and other industries.

HT There is a **real difference** between the mean carbon monoxide concentrations in the city centre and the park because the difference between the mean values is a lot bigger than the range of each set of data. If the difference between the mean values had been smaller than the range, there would have been no real difference. The result would have been insignificant and the data would not support the theory.

Chemicals

Elements are the 'building blocks' of *all* m
There are more than 100 elements and eac
made up of very tiny particles called **atoms**
atoms of a particular element are the **same**
unique to that element.

Each element is represented by a different **chem**
symbol, e.g. C for carbon, O for oxygen and Fe for ir

Atoms can join together to form bigger building blocks, called **molecules**.

Compounds are formed when the atoms of **two or more different elements** are **chemically combined**. The properties of a compound are very different to the properties of the individual elements it is made from.

Chemical symbols and numbers are used to write **formulae**. Formulae show:

- the different elements that make up a compound
- the number of atoms of each different element in one molecule.

Example

A water molecule, H_2O:

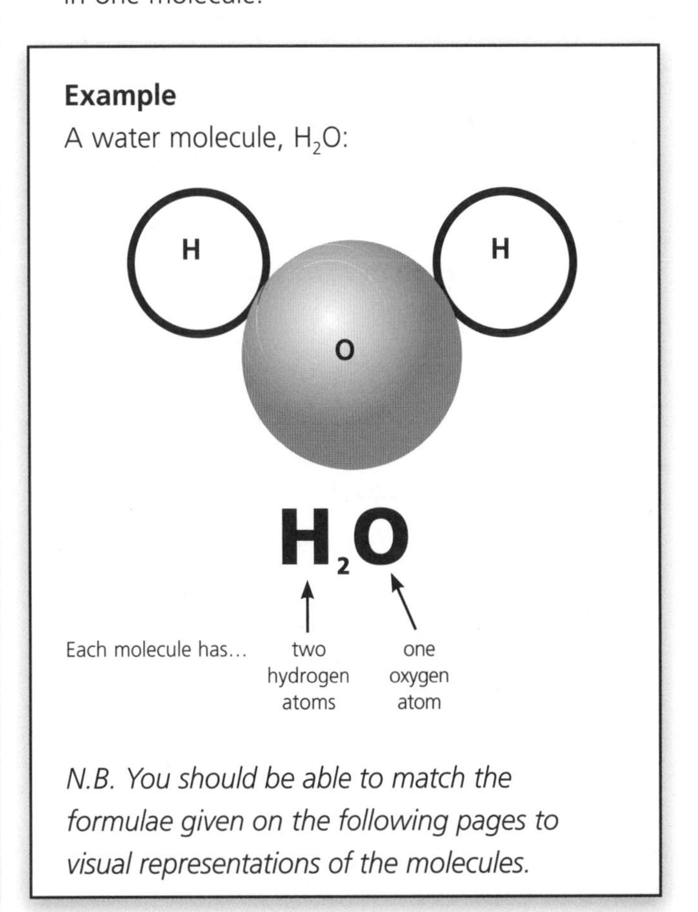

N.B. You should be able to match the formulae given on the following pages to visual representations of the molecules.

...l Change

...hemical reaction new substances are ...om old ones. This is because the atoms in ...ants (starting substances) are rearranged in ...y:

- ...ed atoms may be separated
- ...eparate atoms may be joined
- Joined atoms may be separated and then joined again in different ways.

These chemical changes are **not** easily reversible. You can show what happens during a chemical reaction by using a word equation. The reactants are on one side of the equation and the **products** (newly formed chemicals) are on the other.

Oxidation and Reduction

An example of **oxidation** is a chemical reaction that occurs when oxygen joins with an element or compound. An example of **reduction** is when oxygen is lost from a substance.

Combustion

Combustion is a chemical reaction that occurs when fuels burn, releasing energy as heat. For combustion to take place, **oxygen** must be present. Combustion is an example of oxidation.

Coal is a fossil fuel that consists mainly of carbon. The following equation shows what happens when coal is burned in air:

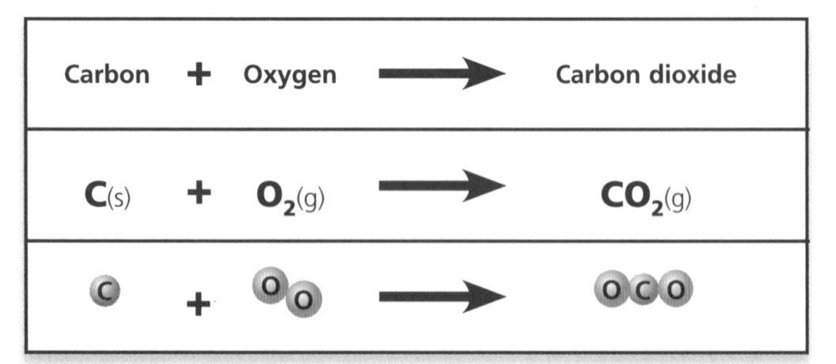

Carbon	+	Oxygen	→	Carbon dioxide
$C_{(s)}$	+	$O_{2(g)}$	→	$CO_{2(g)}$
C	+	O O	→	O C O

This equation tells us that one atom of carbon (solid) and one molecule of oxygen (gas) produce one molecule of carbon dioxide (gas).

No atoms are lost or produced during a chemical reaction. So, there will **always** be the same number of atoms on each side of the equation, therefore conserving mass. This means there will always be some pollutants formed during the combustion of fuels.

HT The conservation of atoms during combustion reactions has implications for air quality since some atoms in the fuel may react to give products that are pollutants, e.g. carbon monoxide or sulfur dioxide.

Burning Fossil Fuels

Many of the pollutants in the atmosphere are produced through the combustion of fossil fuels, e.g. in power stations, cars, aeroplanes, etc.

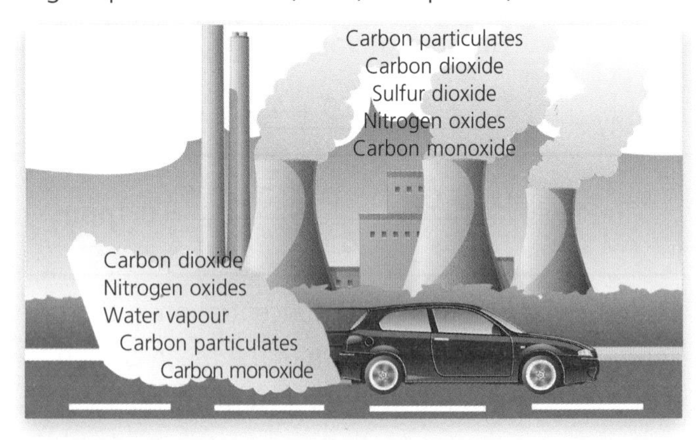

Complete Combustion

Fossil fuels, such as petrol, diesel fuel, natural gas and fuel oil, consist mainly of compounds called **hydrocarbons**. A hydrocarbon contains *only* **hydrogen** atoms and **carbon** atoms. So, when it is burned in air, **carbon dioxide** and **water** (hydrogen oxide) are produced. This is called **complete combustion**. Remember, carbon dioxide is a pollutant!

If the fuel burns in pure oxygen, the reaction is more rapid than when it burns in air.

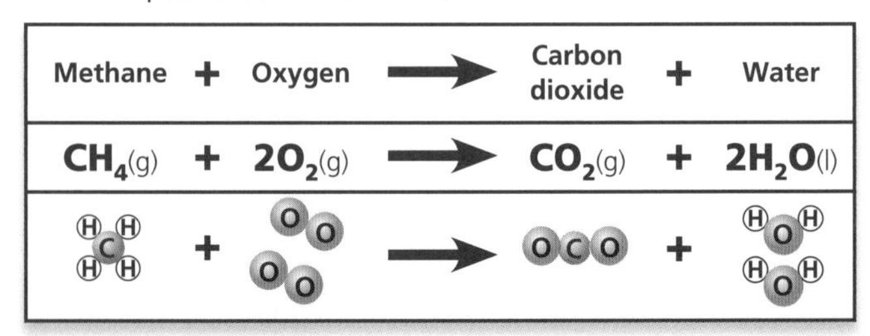

Methane	+	Oxygen	→	Carbon dioxide	+	Water
$CH_{4(g)}$	+	$2O_{2(g)}$	→	$CO_{2(g)}$	+	$2H_2O_{(l)}$
H H C H H	+	O O / O O	→	O C O	+	H O H / H O H

Incomplete Combustion

If a fuel is burned and there is not enough oxygen in the air, **carbon particulates** (**C**) or **carbon monoxide** (**CO**) may be produced. This is called **incomplete combustion**.

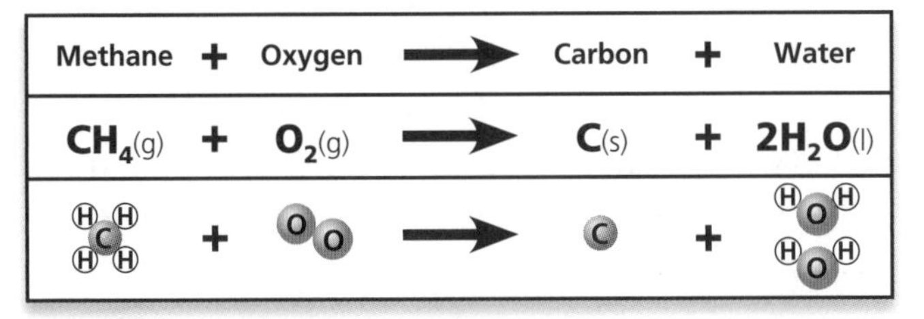

Methane	+	Oxygen	→	Carbon	+	Water
CH_4(g)	+	O_2(g)	→	C(s)	+	$2H_2O$(l)

Methane	+	Oxygen	→	Carbon monoxide	+	Water
$2CH_4$(g)	+	$3O_2$(g)	→	2CO(g)	+	$4H_2O$(l)

Incomplete combustion occurs in car engines, so exhaust emissions contain carbon particulates and carbon monoxide, as well as carbon dioxide.

Many samples of coal contain sulfur, so sulfur dioxide is released into the atmosphere when they are burned.

Sulfur	+	Oxygen	→	Sulfur dioxide
S(s)	+	O_2(g)	→	SO_2(g)

During the combustion of fuels, high temperatures (e.g. in a car engine or power station) can cause **nitrogen** in the atmosphere to react with **oxygen** and produce **nitrogen oxides**.

HT The nitrogen oxides are formed in two steps:

1. The nitrogen reacts with oxygen to form nitrogen monoxide.

Nitrogen	+	Oxygen	→	Nitrogen monoxide
N_2(g)	+	O_2(g)	→	2NO(g)

HT 2. Nitrogen monoxide is then **oxidised** to produce **nitrogen dioxide**.

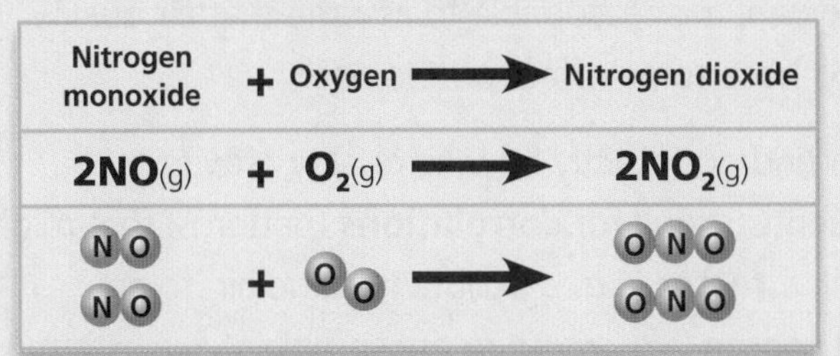

Nitrogen monoxide	+	Oxygen	→	Nitrogen dioxide
2NO(g)	+	O_2(g)	→	$2NO_2$(g)

When NO and NO_2 occur together they are called oxides of nitrogen and they are written as $\mathbf{NO_x}$.

What Happens to Pollutants?

Once pollutants have been released into the atmosphere, they cannot just disappear; they have to go somewhere. This is when they can start causing **problems** for the environment.

Carbon particulates are **deposited** on surfaces such as stone buildings, making them dirty. The appearance of many beautiful old buildings has been changed owing to this.

Some **carbon dioxide** is removed by natural processes; it is needed by plants for **photosynthesis** and some also **dissolves** in rainwater and seawater, where it reacts with other chemicals in the water.

However, because we are producing **too much** carbon dioxide, not all of it is used up naturally. The rest remains in the atmosphere, so each year the concentration of carbon dioxide in the atmosphere increases.

Because carbon dioxide is a **greenhouse gas** (it traps heat in the atmosphere), the rise in concentration is contributing to **global warming**, which is leading to **climate change**.

Sulfur dioxide and **nitrogen dioxide** dissolve in water to produce **acid rain**. Acid rain can damage trees, erode stonework, corrode metal and upset the pH balance of rivers and lakes. If water is too acidic, plants and animals will die and the whole food chain will be affected.

Identifying Health Hazards

Because humans need to breathe in air to get oxygen, it is reasonable to assume that air quality will have some effect on the body.

To find out exactly how air quality affects us, scientists look for **correlations** (patterns) that might link a **factor** (e.g. a pollutant in the air) to an **outcome** (e.g. a respiratory complaint like asthma).

Example
We now know that **pollen** in the air causes **hay fever** in people who have a pollen **allergy**.

However, to reach this conclusion, scientists had to look at thousands of medical records. The data showed that most cases of hay fever occurred in the summer months when pollen counts were high.

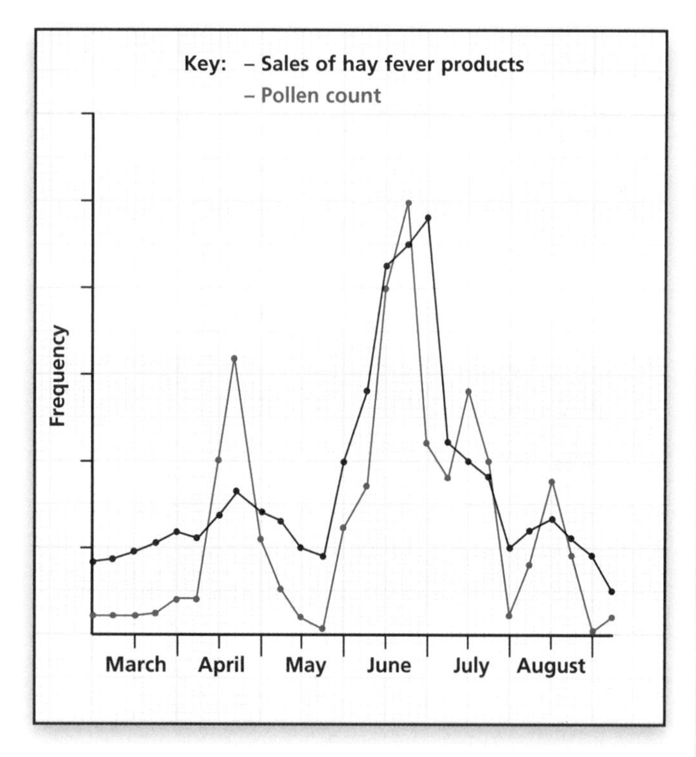

This correlation suggested that pollen **might** cause hay fever. However, it did not provide conclusive evidence because there were lots of other variables that could have influenced the outcome, e.g. temperature, humidity, other pollutants, etc.

Further investigations, in the form of **skin tests**, were carried out to find out how pollen can affect health.

Pollen was collected in spore traps. The pollen was then stuck to the skin of volunteers using plasters.

In some volunteers the skin became red and inflamed, indicating an **allergic reaction**. The results showed that people with a pollen allergy also suffered from hay fever. Those who did not have a pollen allergy did not get hay fever. This provided much stronger evidence of a link between pollen and hay fever.

When these findings were released, other scientists studied the data and repeated the skin test experiments. The fact that the tests always produced the same results proved that they were reliable.

Another condition that is linked to air quality is asthma. However, this example is more complicated. Studies of asthma have shown that when the concentration of NO_2 (nitrogen dioxide) increases in the air, more asthma attacks are triggered.

However, people still have asthma attacks when the levels of nitrogen dioxide are very low. This suggests that although nitrogen dioxide can increase the chance of an asthma attack, it is not the primary **cause**.

There are many factors that can trigger an asthma attack. To fully understand which factors **cause** asthma and which factors may **aggravate** the condition, scientists need to study a large sample of people.

Improving Air Quality

Air pollution affects everyone, so we all have a responsibility to reduce it.

Motor vehicles and power stations that burn fossil fuels are two major sources of atmospheric pollution, so we need to look at how emissions from these sources can be reduced.

Emissions from power stations can be reduced by:

- using less electricity so fewer fossil fuels need to be burned
- removing toxic chemicals before they are burned, e.g. removing the sulfur from natural gas and fuel oil
- using alternative renewable sources of electricity, e.g. solar energy, wind energy and hydroelectric energy, to replace fossil fuels
- using a filter system to remove sulfur dioxide and particulates (carbon and ash) from flue gases before they leave a coal-burning power station's chimney.

HT The sulfur dioxide is removed from flue gases by **wet scrubbing**, using an alkaline slurry or seawater.

During wet scrubbing, the flue gas containing the sulfur dioxide is brought into contact with a slurry of either limestone (calcium carbonate) or lime (calcium oxide) and water. The alkaline slurry is usually sprayed onto the pollutant gas flow, where the sulfur dioxide is absorbed onto its surface. The absorbed sulfur dioxide is then converted to calcium sulfate.

During seawater scrubbing, flue gases are washed with seawater, which dissolves the sulfur oxides. The acidic dissolved sulfur oxides will react with alkalis, such as hydrogencarbonates present in the seawater, to produce sulfate salts.

Emissions from motor vehicles can be reduced by:

- buying a car with a modern engine that is more efficient and burns less fuel
- buying a hybrid car, which uses electric power in the city centre and can then switch to running on petrol for longer journeys
- using a low-sulfur fuel (readily available) to reduce the amount of sulfur dioxide released
- converting the engine to run on biodiesel, which is a renewable fuel
- using public transport to reduce the number of vehicles on the road
- making sure cars are fitted with **catalytic converters**, which reduce the amount of carbon monoxide and nitrogen monoxide emitted.

The reactions that occur in a catalytic converter are:

- Carbon monoxide is oxidised to carbon dioxide by gaining oxygen.

Carbon monoxide	+	Oxygen	→	Carbon dioxide
$2CO_{(g)}$	+	$O_{2(g)}$	→	$2CO_{2(g)}$
CO CO	+	O O	→	OCO OCO

- Nitrogen monoxide is reduced to nitrogen by losing oxygen.

Nitrogen monoxide	+	Carbon monoxide	→	Nitrogen	+	Carbon dioxide
$2NO_{(g)}$	+	$2CO_{(g)}$	→	$N_{2(g)}$	+	$2CO_{2(g)}$
NO NO	+	CO CO	→	N N		OCO OCO

The only way of reducing carbon dioxide emissions is to burn fewer fossil fuels.

Global Choices

In 1997 there was an international meeting about climate change in **Kyoto**, Japan. People from many nations agreed to reduce carbon dioxide emissions, and targets were set for individual countries. The governments of the countries are required to take appropriate measures to meet the targets.

National Choices

Here are some of the initiatives that are helping the UK meet its target:

- Setting legal limits for vehicle exhaust emissions, which are enforced by statutory mot (ministry of transport) tests
- Making catalytic converters compulsory on new vehicles
- Using subsidies (grants) or reduced taxes to encourage power companies to use 'cleaner' fuels
- Introducing a car tax system that encourages drivers to buy smaller cars with smaller engines
- Encouraging investment in non-polluting renewable energy such as wind and solar energy.

These initiatives will impact on many areas of science and industry. For example, when new cars are developed, the technology used must meet all the legal requirements.

Some governments are concerned that steps taken to reduce carbon dioxide emissions will result in a decline in manufacturing and production, employment and the national economy.

Local Choices

Many local authorities are trying to encourage us to make environmentally friendly choices by providing:

- doorstep collections of paper, bottles, metals and plastics for recycling
- regular bus and train services
- electric trams (in some cities)
- congestion charges
- 'park and ride' schemes
- cycle paths and cycle parks.

Personal Choices

It is clear that the **choices** we make as **individuals** affect the amount of pollution in the air.

Using less energy in the home reduces the demand for energy from power stations, e.g. turning off televisions and not leaving them on standby.

Making sure your car is energy efficient and has a catalytic converter, or choosing an alternative mode of transport (e.g. a bicycle), cuts down on vehicle emissions.

Recycling materials like paper, bottles, metals and plastics helps to conserve natural resources but also saves energy, e.g. it takes about 95% less energy to recycle an aluminium can than to make a new one.

There are other benefits to the 'green' options too. For example, walking and cycling instead of travelling by car help to keep us fit!

HT When making national choices, the benefits and needs must be weighed up against all the problems. This could result in different countries making different decisions. For example, there are benefits and problems when using biofuels and electricity to fuel cars.

Fuel	Benefits	Problems
Biofuels	Renewable source that is carbon zero	• A lot of land is needed to grow biomass • Transportation of biomass to the generator
Electricity	No exhaust fumes and a quiet engine	• Lots of charging stations needed • Length of time to recharge batteries • Battery life • Cost

Module C2 (Material Choices)

We use materials for a variety of different functions every day. Materials are often selected for a job because of the properties that they possess. This module looks at:

- the properties and structure of materials
- how polymers are created
- how the properties of materials can be altered
- the importance of nanotechnology.

Natural and Synthetic Materials

The materials that we use are chemicals, or mixtures of chemicals, and include metals, polymers and ceramics. Some materials can be made or obtained from living things, e.g. cotton (plant), paper (wood), silk (a silk worm) or wool (sheep). Synthetic materials, produced by chemical synthesis, can be made as alternatives to these. The raw materials may be taken from the Earth's crust.

Crude Oil

When extracted, crude oil is a thick, black, sticky liquid. It contains mainly **hydrocarbons**, which are chain molecules containing only hydrogen and carbon atoms.

Different hydrocarbons have different boiling points because their molecular chains are different lengths. The strength of the forces between the hydrocarbon molecules increases as the length of the molecule increases. More energy is needed to break the forces between the molecules in the liquid so that they can move freely as a gas. Therefore, larger molecules have higher boiling points. This means that hydrocarbons can be separated by fractional distillation into different parts or **fractions** (groups of hydrocarbons with similar lengths).

The petrochemical industry refines naturally occurring crude oil to produce fuels, lubricants and raw materials for chemical synthesis. Only a small proportion of crude oil is used in chemical synthesis; most of it is used for fuels.

Properties and Uses of Materials

Examples

Unvulcanised Rubbers	
Properties: • Low tensile strength • Soft • Flexible / elastic	**Uses:** • Erasers • Rubber bands
Vulcanised Rubbers	
Properties: • High tensile strength • Hard • Flexible / elastic	**Uses:** • Car tyres • Conveyor belts • Shock absorbers
Plastic – Polythene	
Properties: • Light • Flexible • Easily moulded	**Uses:** • Plastic bags • Cling wrap • Water pipes
Plastic – Polystyrene	
Properties: • Light • Insulation properties • Water resistant	**Uses:** • Meat trays • Egg cartons • Coffee cups • Packaging
Synthetic Fibres – Nylon	
Properties: • Lightweight • Tough • Blocks ultraviolet light	**Uses** • Clothing • Climbing ropes
Synthetic Fibres – Polyester	
Properties: • Lightweight • Waterproof • Tough	**Uses:** • Clothing • Bottles

Properties and Uses of Materials (Cont.)

The properties of the materials used will affect the effectiveness of the end product, so manufacturers always test and assess them carefully beforehand.

Example

A supermarket needs to produce carrier bags. It can use either polythene or biodegradable plastic.

One factor that will determine the supermarket's choice of material is strength, so it carries out the following investigation: a 2cm × 20cm strip of each type of plastic is placed in a clamp. (Each strip used is exactly the same size to ensure a fair test). Weights are then gradually attached to the bottom of each strip to find the total weight it can support before breaking. The experiment is repeated a number of times to ensure the results are reliable.

Measurement	Maximum Weight (N)	
	Polythene	Biodegradable Plastic
1	25.45	19.80
2	25.50	19.75
3	25.40	19.80
4	52.50	19.85
5	25.50	19.90

N.B. 1kg weighs 10N.

When analysing data like this, look to see if any values stand out as being unusual, i.e. they look like **outliers**. In the data collected for polythene the fourth measurement is an outlier.

HT The outlier is likely to have been caused by human error, e.g. the investigator writing down the measurement incorrectly. It is discounted.

The range (or span) of each set of data is from the lowest value to the highest value. The **true value** of the measured quantity is likely to lie within this range. Calculating the mean of a set of data helps to overcome any small variations and obtain a best estimate for the true value of the measured quantity.

Mean Maximum Weight for Polythene $= \frac{101.85}{4}$ **= 25.46N**

Mean Maximum Weight for Biodegradable Plastic $= \frac{99.10}{5}$ **= 19.82N**

This data shows that polythene can support more weight than the biodegradable plastic before breaking. In terms of strength, this makes polythene the most suitable material from which to make carrier bags.

HT On average, the polythene strips can support a weight of 25.46N before breaking, whereas the biodegradable strips can only support 19.82N. Both sets of data show that the range lies within ±0.06N or ±0.08N. This means that, at worst, the polythene strips may break at 25.40N and, at best, the biodegradable strips at 19.90N. This still leaves a difference of 5.50N (25.40 – 19.90), which in terms of strength makes polythene the most suitable material to make the bag from.

However, there are lots of other considerations the supermarket must take into account before making its final decision.

HT Other considerations the supermarket must take into account include:

- being certain that the data can be reproduced
- cost
- biodegradability
- whether the bags are waterproof or not.

Polymerisation

Polymerisation is an important chemical process in which small hydrocarbon molecules, called **monomers**, are joined together to make very long molecules called **polymers**:

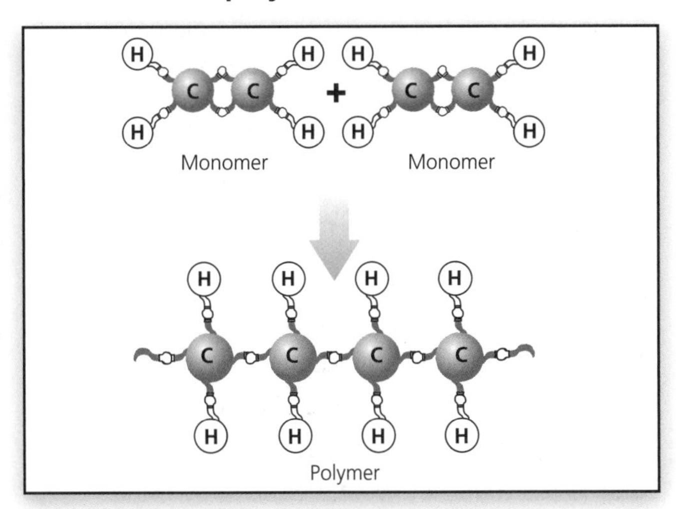

In the example below, the resulting long-chain molecule polymer is polyethene, often called polythene.

$$\begin{array}{ccc} \text{H} \quad \text{H} & & \text{H} \quad \text{H} \\ \diagdown \; \diagup & & \diagdown \; \diagup \\ \text{C}=\text{C} & + & \text{C}=\text{C} \\ \diagup \; \diagdown & & \diagup \; \diagdown \\ \text{H} \quad \text{H} & & \text{H} \quad \text{H} \end{array} \;\; + \text{ many more} \;\longrightarrow\; \left[\begin{array}{cccc} \text{H} & \text{H} & \text{H} & \text{H} \\ | & | & | & | \\ -\text{C} - & \text{C} - & \text{C} - & \text{C}- \\ | & | & | & | \\ \text{H} & \text{H} & \text{H} & \text{H} \end{array}\right]_n$$

Ethene monomers

Polyethene polymer when n = a large number

Remember that during a chemical reaction the number of atoms of each element in the products must be the same as in the reactants. Count the atoms!

Using Polymerisation

Polymerisation can be used to create a wide range of different materials that have different properties and therefore can be used for different purposes.

Many traditional (natural) materials have been replaced by polymers because of their superior properties.

Polymer	Monomer	Use	Traditional Material	Reason
Polyethene	Ethene	Carrier bags	Paper	Stronger; waterproof
Polychloroethene PVC	Chloroethene	Window frames	Wood	Unreactive; does not rot

Molecular Structure of Materials

The properties of solid materials depend on how the particles they are made from are arranged and held together.

Natural rubber is very flexible. It consists of a tangled mass of long-chain molecules. Although the atoms in each molecule are held together by strong covalent bonds, there are very weak forces between the molecules so they can easily slide over one another, allowing the material to stretch.

Rubber has a low melting point as little energy is needed to separate the molecules.

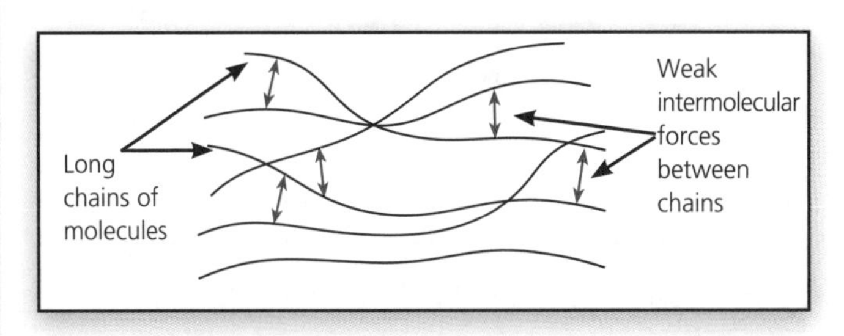

Materials with strong forces between the molecules (covalent bonds or cross-linking bridges) have high melting points as lots of energy is needed to separate them.

As the molecules cannot slide over one another, these materials are rigid and cannot be stretched.

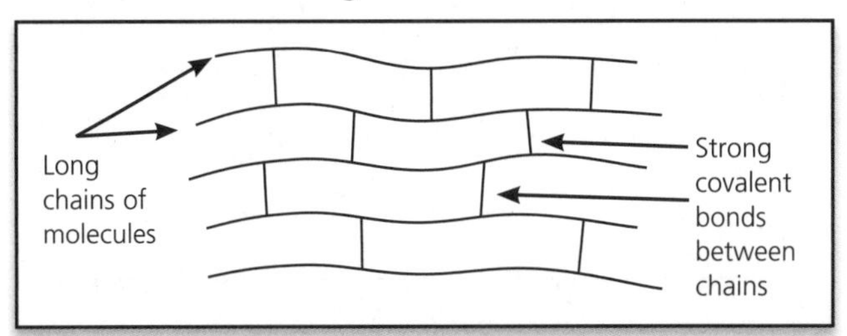

Modifications in Polymers

Modifications can produce changes to the properties of polymers. These modifications can include:

- **increasing the chain length** – longer molecules are stronger than shorter ones
- **cross-linking** – cross-links are formed by atoms bonding between the polymer molecules, so they are no longer able to move. This makes for a harder material. An example of this is **vulcanisation**, when sulfur atoms form cross-links between rubber molecules. Vulcanised rubber is used to make car tyres and conveyor belts.
- **plasticisers –** adding plasticisers makes a polymer softer and more flexible. A plasticiser is a small molecule that sits between the molecules and forces the chains further apart. The forces between the chains are, therefore, weaker and so the molecules can move more easily. Plasticised PVC is used to make children's toys.

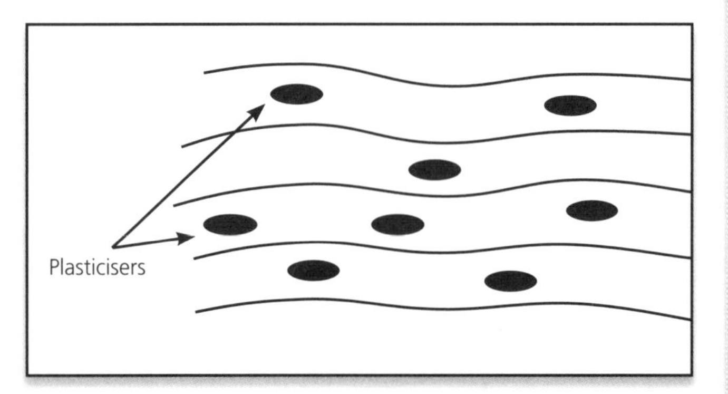

HT A polymer can also be modified by packing the molecules more closely together to form a **crystalline polymer**. The intermolecular forces are slightly stronger so the polymer is stronger, more dense and has a slightly higher melting point.

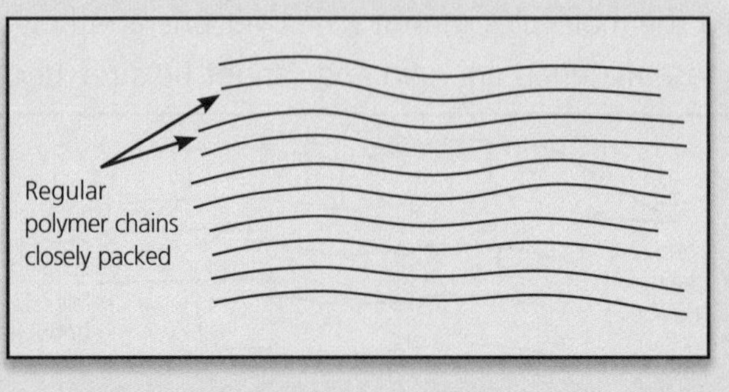

Nanotechnology

Nanoscience refers to the study of materials that are 1–100 nanometres in size, which is roughly the size of a few atoms. One nanometre is 0.000 000 001m (one billionth of a metre) and is written 1nm or 1×10^{-9}m. (A human hair is around 20 000nm in diameter and a microbe is around 200nm in diameter.)

Nanotechnology is the science of building things on a very tiny scale. It is the understanding and control of matter at dimensions between approximately 1 and 100 nanometres. Nanotechnology is a growing industry.

Nanoscale materials are not new. Naturally occurring materials, such as liposomes and seaspray, have always existed. However, it was not until the early 1980s that they were identified by using scanning tunnelling microscopes.

Nanoscale materials are designed to do specific jobs. For example, nanoparticles of titanium dioxide are added to sunscreen as they are very efficient at absorbing ultraviolet radiation. They are also being developed for use within medicine and dentistry, as well as in the car industry and product-specific catalysts.

Some nanoscale materials are formed accidentally as a result of other chemical reactions, e.g. the smallest particles from the combustion of fuels.

Some Examples of Uses of Nanomaterials

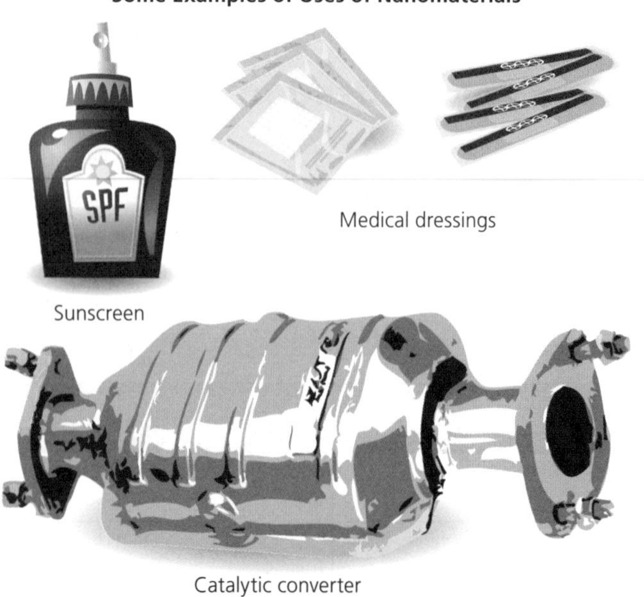

Properties of Nanoscale Materials

Nanotubes and buckyballs are nanoscale objects made of carbon atoms. They have been used in the manufacture of sports equipment, such as badminton rackets, for many years. In the future they are likely to play an important role in electronic systems.

A Buckyball

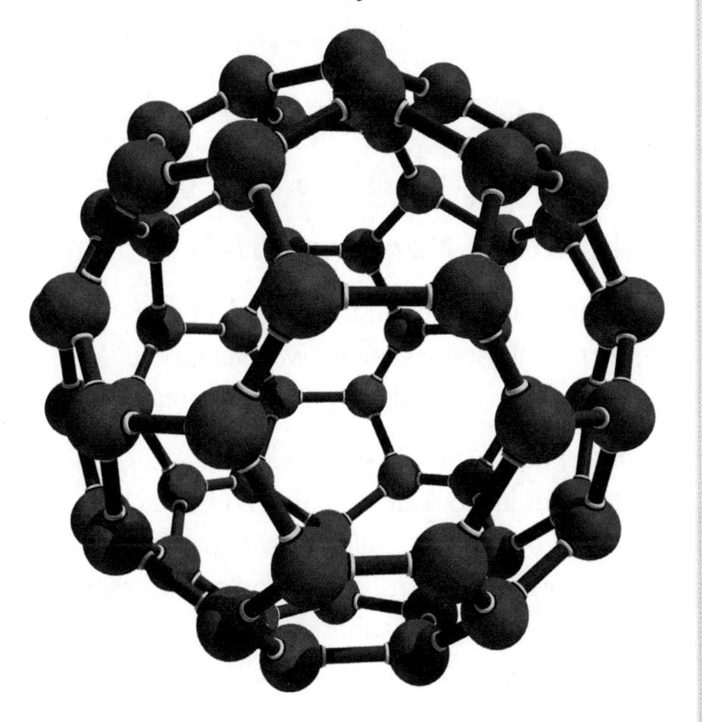

Nanoscale particles have different properties to larger particles of the same material. For example:

- nanoparticle electrons can move through an insulating layer of atoms
- nanoparticles are more sensitive to light, heat and magnetism
- nanoparticles in sunscreens and cosmetics absorb and reflect the harmful ultraviolet rays in sunlight
- nanoparticles can be added to glass to repel water and keep windows clean.

Many of these properties can be explained by the much larger surface area of the nanoparticles compared to their volume.

This means that nanoparticles can be used to modify the properties of existing materials, such as polymers, to make them stronger, stiffer, lighter, etc.

A Magnified Representation of Iron Atoms (Nanoparticles) in a Ring Around Some Surface State Electrons

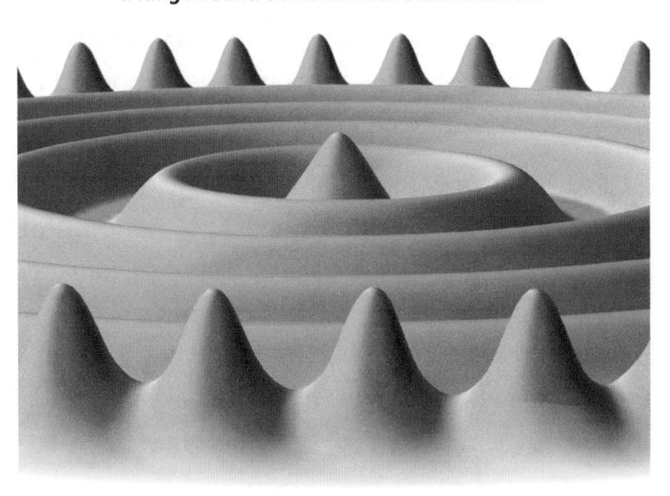

Antibacterial Fibres

In recent years a greater awareness of contact disease transmission and personal hygiene has led to the development of antibacterial fibres to protect wearers against the spread of bacteria and diseases. Researchers have found that silver nanoparticles can destroy many types of bacteria.

Using this knowledge, scientists have developed antibacterial fibres containing silver nanoparticles that are woven into textiles and used to make clothes. Many leading sports-clothing manufacturers now use this silver oxide fibre technology. One of the special features of the clothing is that it has antibacterial properties, which keep the garments fresh.

Nanotechnology in Sports

It is not just the sportswear industry that is using nanotechnology. Nanotechnology is being applied to many sports, e.g. tennis and golf.

Golf clubs are now much lighter, stronger and more efficient than they used to be, thanks to **nanometal** coatings. Nanometals have a crystalline structure and, although they are hundreds of times smaller than traditional metals, they are four times stronger. Golf balls are now treated with nanoscale particles that allow them to travel in straighter lines.

Leading manufacturers of sports equipment have also started adding nanoscale silicon dioxide crystals to tennis rackets. The resulting polymer gives increased performance, without changing the weight.

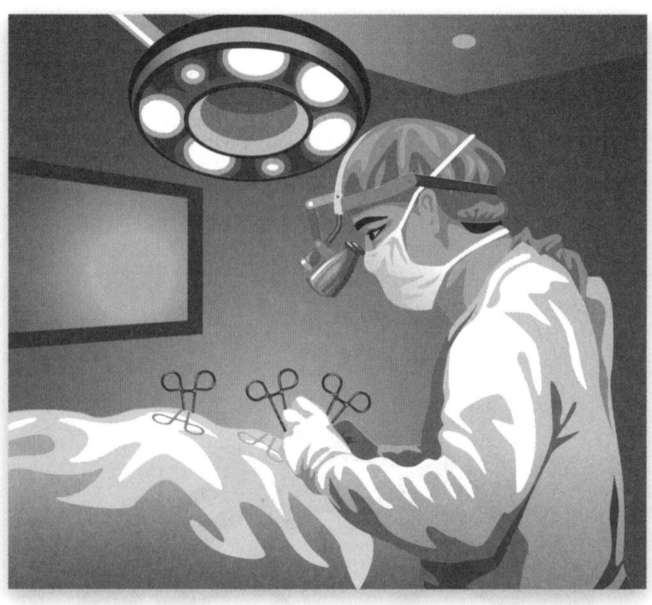

Evaluating Nanomaterials

Nanotechnology is still in the early stages of development. New materials, with very useful properties, are being developed all the time.

Nanotechnology has a variety of potential applications in biomedical, optical and electronic fields. For example, it could be used to create secure communication systems, detect and eradicate small tumours, help in the diagnosis of diseases and help in the development of microscopic surgery that would not leave scars.

It is important to remember that nanoparticles can be dangerous in certain circumstances. For example, nanoparticles in water could be dangerous if they were consumed. Nanoparticles may have other harmful effects on health that are currently not known about.

Some people and organisations are extremely concerned that products with nanoparticles are being introduced before they have been fully tried out and tested. It takes a long time to carry out a full investigation, and any harmful health effects may not be apparent for many years.

Regulations for the development of new techniques and products do exist. A report by the Royal Society suggests that these regulations are adequate to deal with most of the nanotechnology products.

Module C3 (Chemicals in Our Lives: Risks and Benefits)

Britain is a country that has large deposits of valuable minerals, which have been the basis of the chemical industry for more than 200 years. This module looks at:

- the origins of minerals in Britain that contribute to our economic wealth
- plate tectonics
- the importance of salt and where it comes from
- the alkali industry
- making chlorine
- how to make our chemicals safe and sustainable
- life cycle assessment.

The Origins of Mineral Wealth in Britain

Geologists are scientists who study rocks and the processes that formed them. They try to explain the past history of the surface of the Earth by modelling processes that can be observed today.

We know that the Earth's lithosphere (the crust and the upper part of the mantle) is 'cracked' into several large pieces, called **tectonic plates**.

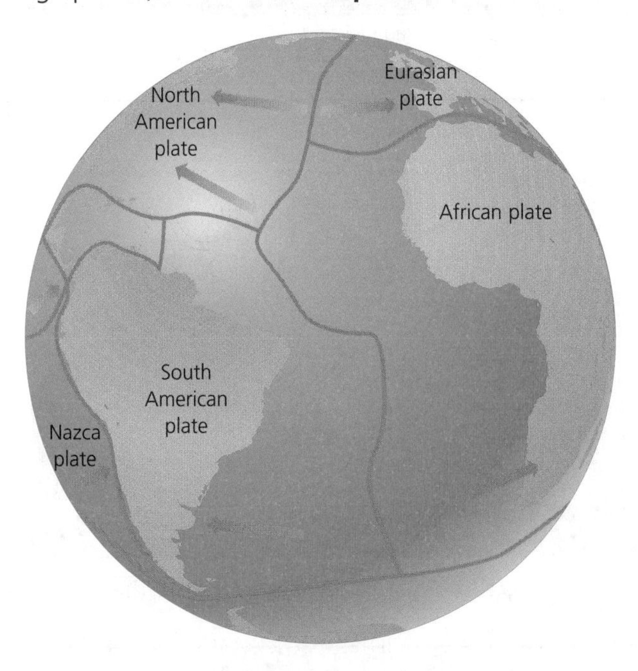

Intense heat, released by radioactive decay deep in the Earth, causes hot molten rock to rise to the surface at the boundary of the plates. This causes the tectonic plates to move very slowly, at speeds of a few centimetres per year.

Geologists use magnetic clues in rocks to track this very slow movement of the plates. They have shown that parts of ancient continents have moved over the surface of the Earth to make up Britain as we know it today. As a result, rocks found in different parts of Britain were formed in different climates.

Over millions of years, a number of processes have led to the formation of valuable resources in Britain, such as coal, limestone and salt. These processes include:

- mountain formation
- erosion
- sedimentation
- dissolving
- evaporation.

These processes form part of the **rock cycle**, which you may have studied during Key Stage 3:

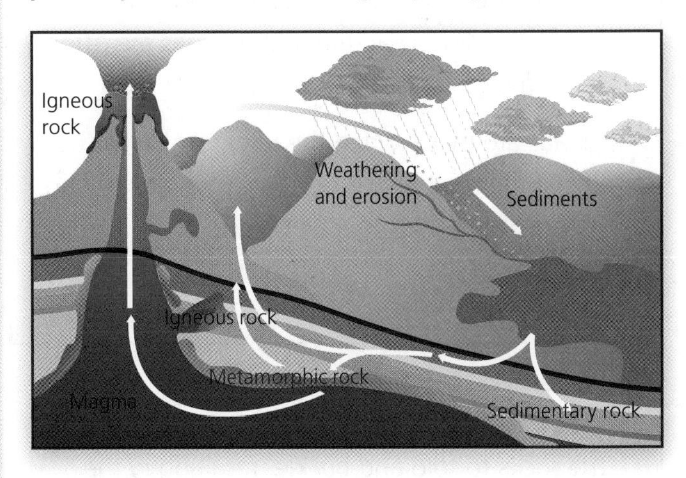

Igneous rocks are formed from molten magma and contain interlocking crystals. Granite and basalt are both igneous rocks.

The formation of sedimentary rocks is described on page 19.

Metamorphic rocks are usually formed from sedimentary rocks subjected to intense heat and pressure. Examples of metamorphic rocks are slate (from shale) and marble (from limestone).

Tectonic Plate Movement

Earthquakes occur at the boundaries of tectonic plates, and mountains are formed when collisions occur between tectonic plates.

1 Plates Slide Past Each Other

When plates slide past each other, huge stresses and strains build up in the crust. These stresses and strains need to be released in order for movement to occur. This 'release' of energy results in an earthquake.

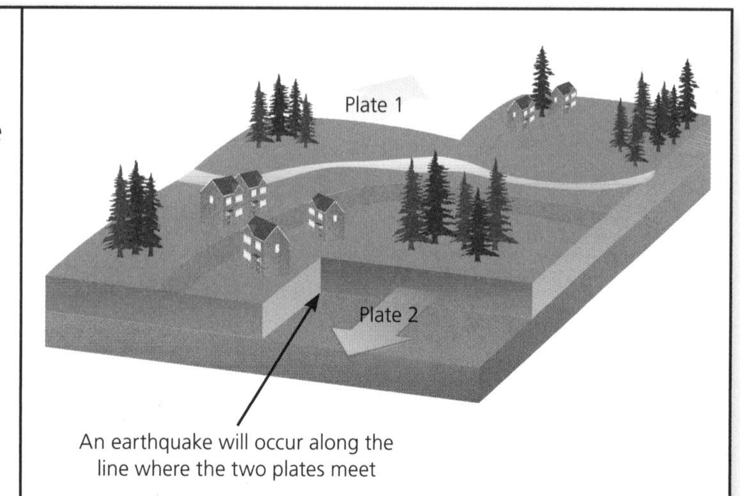

2 Plates Move Away from Each Other

When plates move away from each other, fractures in the crust occur at the boundary. Molten rock rises to the surface, where it solidifies. Mid-ocean mountain ridges are often formed under the ocean this way.

Islands are made when the new rock builds up above the level of the sea. For example, Iceland is part of the Mid-Atlantic Ridge.

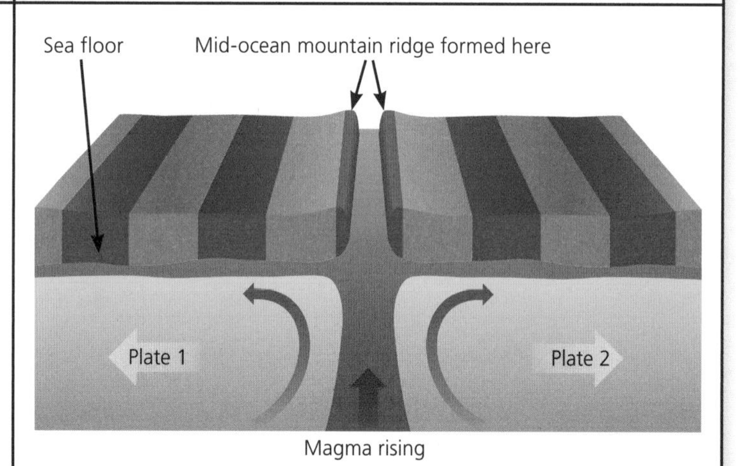

3 Plates Move Towards Each Other

When plates collide, the huge pressures cause the rocks to fold and buckle, resulting in the formation of mountain chains. Sometimes as the plates collide, one is forced under the other and new mountains are made along the plate boundary.

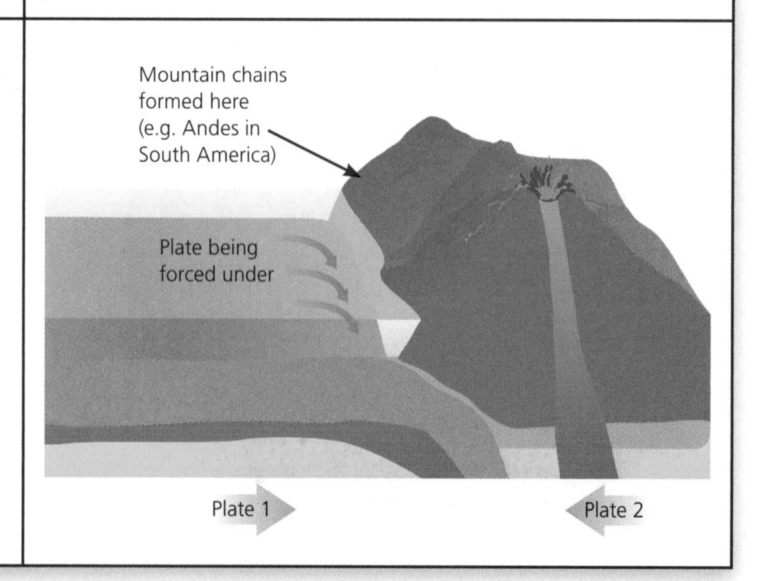

Moving Rocks

As newly formed mountains are exposed to the climate, they are weathered by biological, chemical and physical processes, as shown in the rock cycle. As a result, small fragments of rock are broken off. The fragments are transported, often through rivers, to different places by the process of erosion.

During erosion the fragments are broken down further, into smaller pieces or sediments, as they bump into things. Minerals in the rock, such as salts, dissolve and are carried by the river to different places. Eventually, the river deposits the small sediments on the riverbed or as they enter a lake or the sea. During warm weather water from enclosed lakes **evaporates**, leaving beds of sedimentary **evaporate minerals** including salt crystals.

Organic waste such as leaves, or the skeletons of marine animals, will also be deposited on the river or sea bed. Over millions of years, the layers build up and **sedimentation** occurs. These processes result in the formation of sedimentary rocks such as rock salt, limestone and coal.

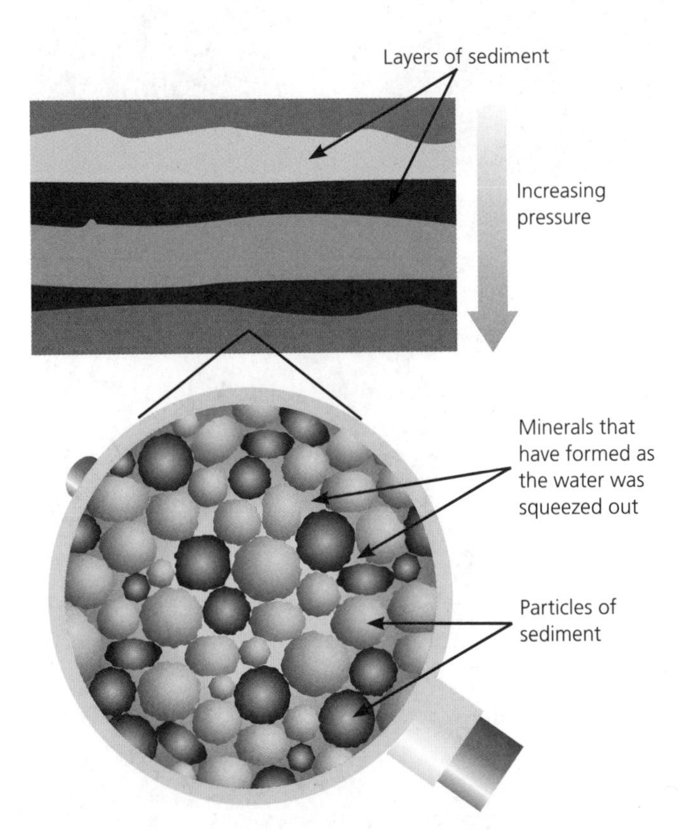

Looking at the Evidence

Today geologists study sedimentary rocks to try to understand how they were formed and where the rocks came from. They look for clues buried in the rocks including:

- fossils
- the presence of shell fragments
- ripples from sea or riverbeds
- the shapes of water-borne grains compared to air-borne grains.

Evidence from Rocks

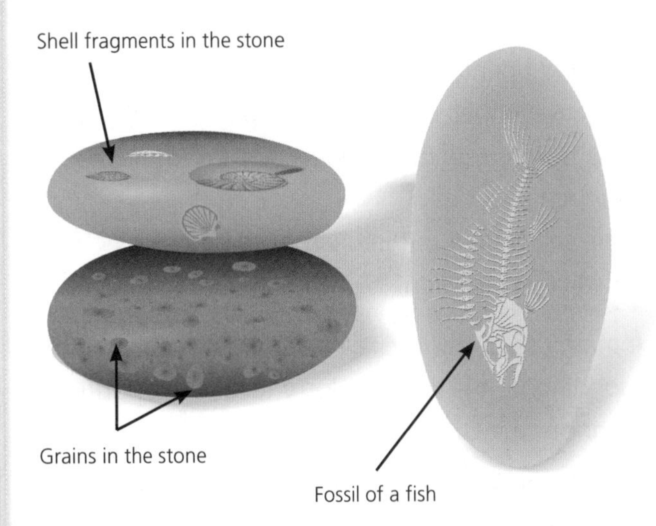

Exploiting our Natural Resources

The chemical industry developed in the north-west of England because resources such as salt, limestone and coal were available locally. This meant that raw materials could be mined and used in one place, rather than having to be transported to a different part of the country. This made economic sense and provided a good range of jobs for local people.

Salt Mining

Salt can be obtained from the sea or from underground salt deposits. Two different processes are used for extracting the salt from underground deposits. The method used may be determined by how the salt is going to be used.

Method 1: Mining

Rock salt is mined in Cheshire, in the north-west of England. The stages in the process are as follows:

1. Explosives are used to blast the exposed layer of rock salt.
2. The rock salt is loaded into a crusher, where it is ground up into small pieces.
3. A conveyor belt transports the salt to the lift shaft.
4. It is transferred into hoppers and taken to the surface.
5. The salt is then put into large storage areas awaiting collection.

The main use of this rock salt is to treat roads during icy conditions. It is taken by lorry to local authorities throughout the UK.

Under the Cheshire countryside, there are more than 120 miles of empty mine tunnels. The salt mines have also left scars on the landscape, especially where large areas of the ground have been dug out.

Method 2: Solution in Water

Salt is soluble in water. Important industrial chemicals such as chlorine and sodium hydroxide are extracted from salt by the electrolysis of brine (salt solution). Salt can be extracted from the ground in solution and piped directly to the electrolysis plant.

The stages in the process are as follows:

1. Holes are drilled into the salt deposits.
2. Explosive or hydraulics may be used to make the holes larger and so make it easier for the water to penetrate the rock salt.
3. Water is pumped into the bores and the salt dissolves.
4. The salt solution is then pumped back to the surface and piped to the processing plant.

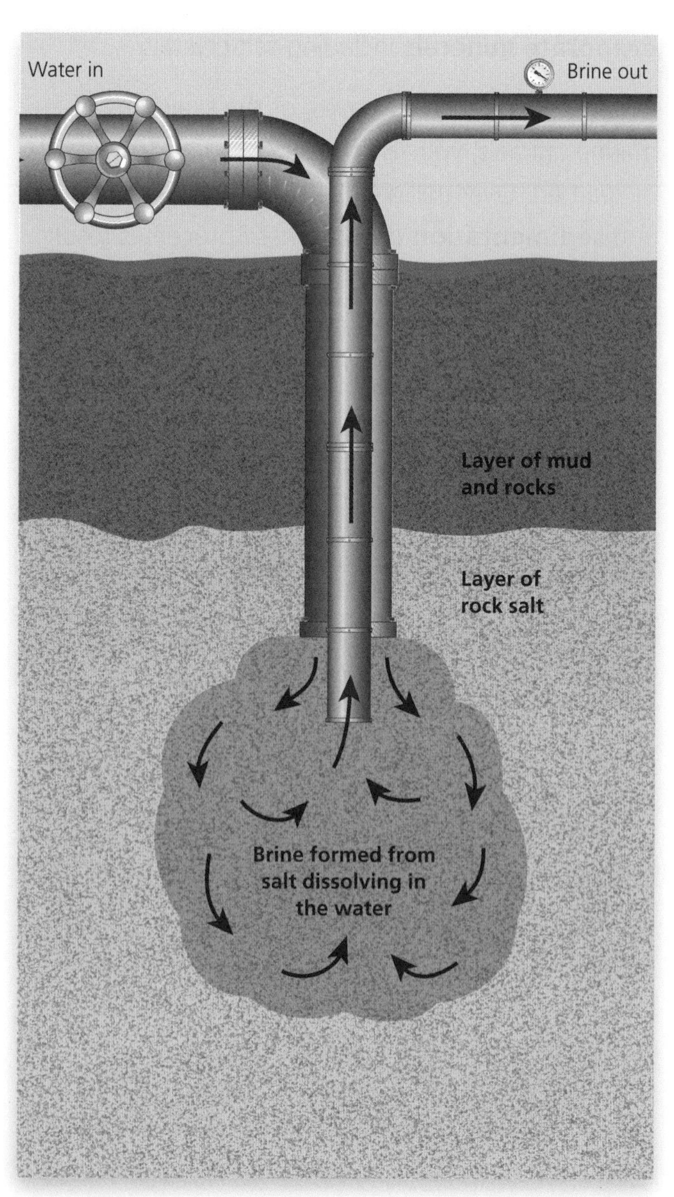

Uses of Salt

Salt (sodium chloride) is a very useful chemical. Here are some of its uses:

A source of chemicals

Treating the roads

Chlorine

Sodium hydroxide

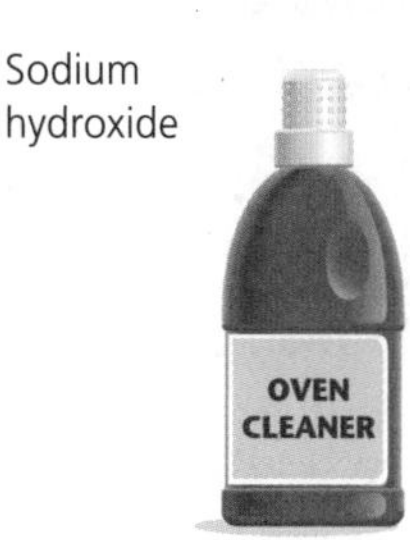

Salt as a Food Additive

Salt is added to food for flavouring and as a preservative. Sodium is present in additives such as monosodium glutamate and sodium bicarbonate. Processed foods, such as meat and bread products, can be high in salt. Processed foods are thought to account for about 75% of the average person's salt intake.

Salt and Health

Salt is an important component of a healthy diet. It is needed to maintain the concentration of body fluids. It helps cells to take up nutrients and plays a crucial role in the transmission of electrical impulses in the nerves.

However, too much salt is not good for you. When the levels of sodium are too high, it causes water to be retained in the body, which means the volume of fluids increases. Some scientists think that this results in high blood pressure, which can increase the risk of heart attacks and strokes.

Government guidelines recommend that adults should eat 6g of salt a day. However, the average intake of salt is between 9g and 10g a day.

Experts estimate that if average consumption was cut to 6g a day, it would prevent 70 000 heart attacks and strokes a year.

Food and the Government

Government departments, such as the Department of Health and the Department for Environment, Food and Rural Affairs, have a role in making sure that our food is safe, healthy and fairly marketed. They also make sure that food producers are acting within the law. The government departments promote healthy eating and aim to minimise illnesses such as food poisoning. They make sure that food labels are clear and that they say exactly what is in the food.

The food labels help people to decide whether or not to buy the product. For example, coeliacs look for labels that say 'gluten free' and vegetarians look to see if the food contains any animal products. (Some foods also state 'suitable for vegetarians'.)

It is important to give the public the most up-to-date information about food safety. In fact, the Food Standards Agency (FSA), an independent government department, was set up to protect public health and consumer interests in relation to food. Government scientists carry out research into food issues such as **genetically modified** (GM) foods.

Sometimes the research findings are controversial and the results are uncertain. Scientists may even disagree about what the results actually mean. Further problems may be encountered from manufacturers who may not want to accept the research findings, as it may not be in their economic interest.

If there is any doubt about food safety, then one of the scientific advisory committees is asked to carry out a risk assessment. It must decide:

- if the food contains any chemicals that could cause harm
- how harmful the chemicals are
- how much of the food must be eaten before it is likely to harm people
- if any groups of people are particularly vulnerable, e.g. the elderly, children, or those suffering from illness.

HT The outcome of a risk assessment is often based on experience gained from people or animals eating the food.

Sometimes the scientific evidence is uncertain and the risk is unknown, in which case the **precautionary principle** is applied. Both experts and the public are consulted before the regulators make a decision about food safety.

Regulators have to weigh up the costs and benefits of any decision, as the priority is to protect public safety and not just let the new foods be mass produced and put on the market.

For example, many people ask questions such as 'Are GM foods safe to eat?'

For many GM foods, scientists simply do not know enough about the science of altering genes, which may lead to health problems in the future. There is also not much data yet on the potential risks to humans, and this is why the precautionary principle is sometimes applied.

The Alkali Industry

Long before industrialisation, alkalis were used in everyday life. Alkalis are very important chemicals as they neutralise acids to make salts. Traditional sources of alkali included burnt wood and stale urine.

Here are some of the uses of alkalis:

- neutralising acidic soil
- producing chemicals that bind natural dyes to cloth
- producing soap
- producing glass.

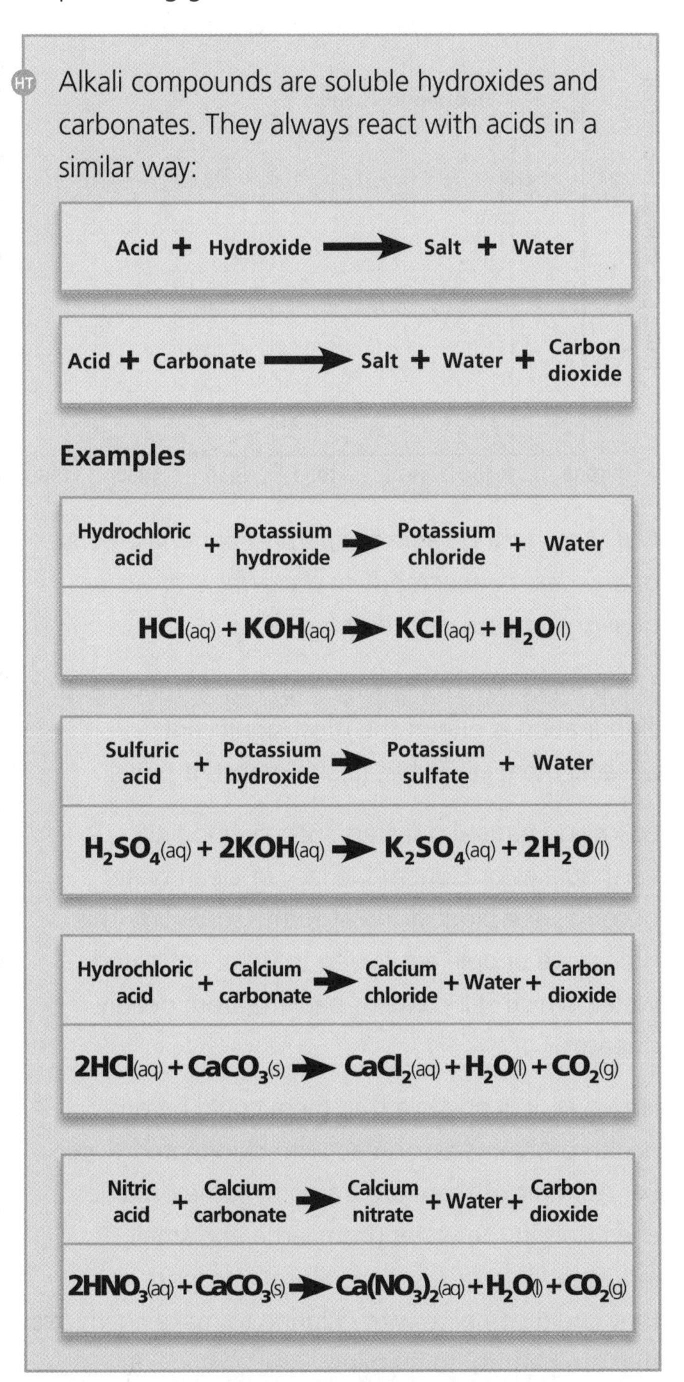

HT Alkali compounds are soluble hydroxides and carbonates. They always react with acids in a similar way:

Acid + Hydroxide → Salt + Water

Acid + Carbonate → Salt + Water + Carbon dioxide

Examples

Hydrochloric acid + Potassium hydroxide → Potassium chloride + Water

$$HCl(aq) + KOH(aq) \rightarrow KCl(aq) + H_2O(l)$$

Sulfuric acid + Potassium hydroxide → Potassium sulfate + Water

$$H_2SO_4(aq) + 2KOH(aq) \rightarrow K_2SO_4(aq) + 2H_2O(l)$$

Hydrochloric acid + Calcium carbonate → Calcium chloride + Water + Carbon dioxide

$$2HCl(aq) + CaCO_3(s) \rightarrow CaCl_2(aq) + H_2O(l) + CO_2(g)$$

Nitric acid + Calcium carbonate → Calcium nitrate + Water + Carbon dioxide

$$2HNO_3(aq) + CaCO_3(s) \rightarrow Ca(NO_3)_2(aq) + H_2O(l) + CO_2(g)$$

With increased industrialisation, and more demand for alkaline-based products, there was a shortage of alkali in the 19th century. As a result, people looked for other ways of processing alkali.

Early processes for manufacturing alkali from salt and limestone caused a lot of pollution. Large volumes of the acidic gas hydrogen chloride were released into the atmosphere and great heaps of waste that slowly released the toxic and foul-smelling gas, hydrogen sulfide, were also formed.

Industrialists have a responsibility to try to minimise the pollution caused by chemical processes. Sometimes the problems can be solved by converting the waste pollutant into a useful chemical. For example, in this case, by dissolving hydrogen chloride gas in water you can make hydrochloric acid:

Hydrogen chloride + Water → Hydrochloric acid

Alternatively, the hydrogen chloride gas could be used to make chlorine gas by oxidising it.

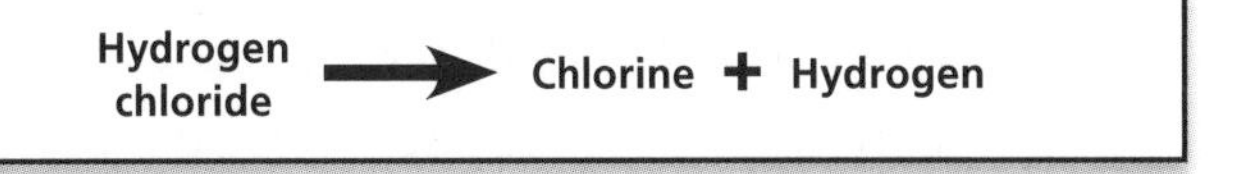

Hydrogen chloride → Chlorine + Hydrogen

Remember that the properties of compounds are different from those of the elements from which they are made. For example, in the reaction above:

- hydrogen chloride is a colourless, acidic gas
- chlorine is a green gas
- hydrogen is a colourless gas.

Making Chlorine

Chlorine is produced by the electrolysis of brine (sodium chloride solution):

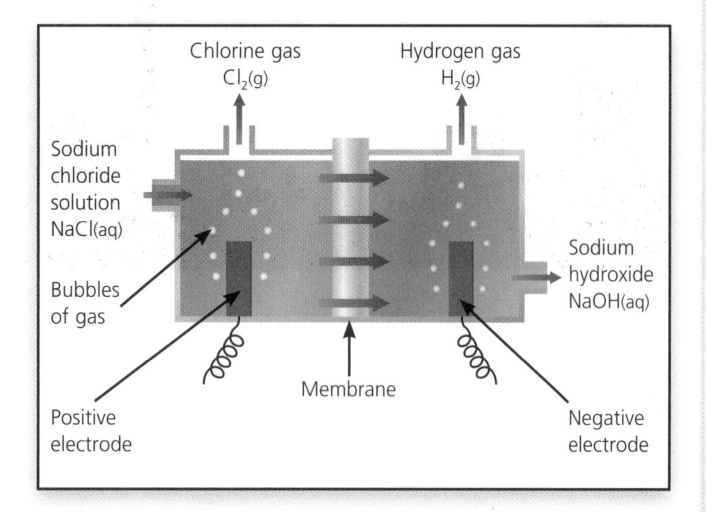

Passing an electric current through the brine causes a chemical change to take place. This forms new products – **chlorine**, **sodium hydroxide** and **hydrogen**. The products have many applications:

- Chlorine is used to kill bacteria in drinking water and swimming pools, and to manufacture hydrochloric acid, disinfectants, bleach and the plastic PVC.
- Sodium hydroxide is used in the manufacture of soap, paper and ceramics.
- Hydrogen is used in the manufacture of ammonia and margarine.

The electrolysis of brine can have an impact on the environment. A major concern is the amount of energy required to carry out electrolysis. A cheap supply of renewable energy is needed.

Water Purification Using Chlorine

Chlorine is added to domestic water supplies to kill any harmful microorganisms that might be present.

Chlorinated drinking water protects against illnesses including:

- typhoid fever
- dysentery
- cholera
- gastroenteritis.

However, chlorinated drinking water will not kill viruses or parasites.

Chlorine was first introduced into drinking water in England in the 1890s and in the USA in 1908.

The graph below shows that, following the introduction of chlorination, there was a decline in the death rate due to typhoid fever in the USA. As more cities across the USA adopted the practice, further reductions were seen until the illness was eliminated in the mid-1940s.

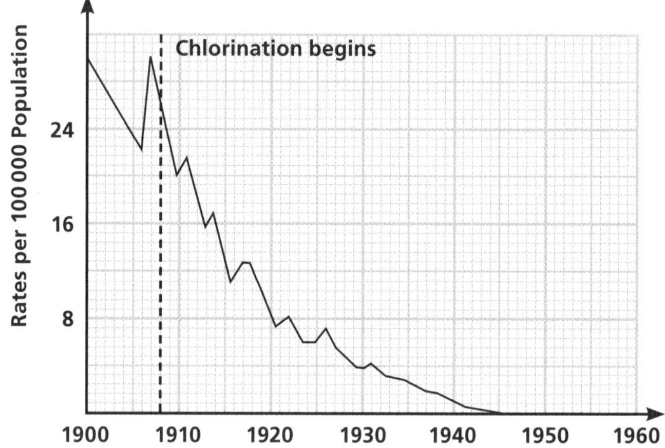

N.B. You must be able to interpret and make sense of any data connected with the effects of water quality that you are presented with.

The purification of drinking water by filtration and chlorination is one of the most significant advancements in public health in recent times.

Following natural disasters such as flooding, earthquakes or tsunamis, a lack of clean water becomes the biggest threat to the survivors. This is because people are forced to drink untreated water, which still contains bacteria from deadly diseases.

However, it is possible that there could be other health problems associated with chlorinated water. For example, if there are traces of some organic chemicals in the water (from large and small industrial enterprises, agriculture, transport, etc.), they could combine with chlorine to make chemicals that are harmful to humans.

Research scientists are continuously investigating these possibilities and offering their advice. From time to time there will be some problems with water quality, such as the scare in North Wales described below.

Water Scare Over

There has been an outbreak of cryptosporidium, a stomach illness that can be caught by drinking infected water. In Anglesey and Gwynedd, there have been 231 reported cases of people catching the bug. As a result, everyone in the North Wales area had to boil their water before consuming it to remove the risk of getting infected.

The infected water is likely to have come from the Llyn Cwellyn reservoir, but health officials have now given North Wales' residents the all-clear; there is no need to continue boiling their water.

However, what remains to be seen is the impact of this scare. General consensus is that people are still wary of their water, and some say they have switched to drinking only bottled water.

Safe and Sustainable Chemicals

Today the pace of scientific and technological development is very fast. New chemicals and materials are being produced continuously. Many materials have useful consumer applications (e.g. in nanotechnology), or they are used in the food industry (e.g. genetically modified foods) or pharmaceutical industry (e.g. for the production of new medicines).

For many new products, scientists have not yet been able to collect enough data to judge whether they are likely to present a risk to the environment and/or human health. Therefore, government departments or individual professionals (e.g. doctors) must decide if the potential benefits outweigh the potential risks.

Given below are some examples where materials were manufactured in large quantities without a potential risk being known.

Thalidomide

In the 1950s, a drug called thalidomide was prescribed to pregnant women to relieve the symptoms of morning sickness. It had been tested on animals and was considered safe to use.

Thousands of babies were subsequently born across the world with serious limb defects and a common factor was found – all the mothers had taken thalidomide in early pregnancy. The drug was withdrawn in 1961. During the initial investigations the drug was never tested on pregnant animals.

Chlorofluorocarbons (CFCs)

In the 1970s, a link was made between CFCs and the destruction of the ozone layer. CFCs had been used in refrigerants and aerosol sprays. Scientists had thought that CFCs were very unreactive molecules, posing no environmental risks.

However, it was discovered that once CFCs are released into the environment, they are carried very large distances into the upper atmosphere where they react with ozone and destroy it.

Potentially, the molecules could be in the environment for 300 years before they react with the ozone.

Polyvinyl Chloride (PVC)

PVC is a polymer that contains carbon, hydrogen and chlorine atoms:

$$H_2C{=}CHCl + H_2C{=}CHCl \xrightarrow{\text{+ many more}} \left[-CH_2-CHCl-CH_2-CHCl-\right]_n$$

Chloroethene monomers — PVC polymer

Plasticisers can be added to the PVC to make it softer and more flexible, so that the range of uses can be expanded. Scientists have now found that the plasticisers can be leached from the plastic into the surroundings, where they may have harmful effects.

Life Cycle Assessment (LCA)

The life cycle of a product has four phases – **Making the material from natural raw materials**, **Manufacture**, **Use** and **Disposal**. The LCA involves examining each of these phases in detail, including the impact on the environment.

Each part of the life cycle of a product is carefully considered and assessed on the amount of energy and materials (including water) that will be used and how materials will be obtained and disposed of. The outcome of the LCA is dependent on several factors, including the use of the end product.

LCAs were introduced in the 1960s to encourage companies to reduce waste and be aware of environmental impact. New laws were put in place to protect the environment; cash incentives were offered to encourage recycling; and in 1996 a tax was introduced to discourage the use of landfill sites.

The purpose of an LCA is to ensure the most **sustainable** method is used, which means meeting the needs of today's society, whilst allowing for the needs of future generations.

The flow diagram below shows what is being assessed in each stage of an LCA.

Making the material from natural raw materials: Natural raw materials, water and energy needed to make the starting material. Environmental impact from obtaining the natural raw materials.

↓

Manufacture: Resources (including water) and energy needed to make the product. Environmental impact of making the product from the material.

↓

Use: Energy needed to use the product, e.g. fuel and electricity. Energy and chemicals (including water) needed to maintain the product. Environmental impact of using the product.

↓

Disposal: Energy needed to dispose of the product. Environmental impact of landfill, incineration and recycling.

Materials and Their Functions

Different materials can often be used to perform the same job. For example, disposable nappies are made from cellulose fibres, a super-absorbent polymer and fluff pulp, whilst re-usable nappies are made from cloth. Disposable nappies may be more convenient but in a life cycle assessment which one is better for the environment?

The results of one study are shown below.

Impact Per Baby, Per Year	Re-usable Nappies	Disposable Nappies
Energy needed to produce product	2532MJ	8900MJ
Waste water	12.4m^3	28m^3
Raw materials used	29kg	569kg
Domestic solid waste produced	4kg	361kg

The evidence here shows that using re-usable nappies consumes less energy, water and natural resources, whilst also producing less waste. This would suggest that people should be encouraged to use re-usable nappies.

Since 2003 it has been government policy to encourage parents to reduce the number of disposable nappies they use.

The same material can be used to perform different jobs. For example, Teflon® (polytetrafluoroethene), which was accidentally discovered in 1938 by Roy Plunkett, can be used in:

- gaskets and valves
- insulation
- non-stick saucepans
- dentures.

Teflon® is **chemically inert** and temperature resistant, and there is also little impact on the environment when it is disposed of in landfill sites.

1 **(a)** The Earth was formed 4.6 billion years ago. The planet's atmosphere at that time was very different to today's atmosphere. Explain how it changed by filling in the empty boxes with the letters to put the stages in the correct order. The first one has been done for you. **[3]**

A Water vapour condensed to form oceans.

B Volcanic activity released carbon dioxide.

C Nitrogen was made as gases reacted with oxygen.

D Green plants evolved.

E Carbon dioxide levels decreased as oxygen was formed.

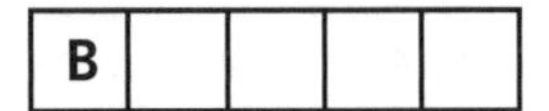

(b) Which pie chart represents the current composition of gases in the Earth's atmosphere? Put a tick (✓) in the box next to the correct answer. **[1]**

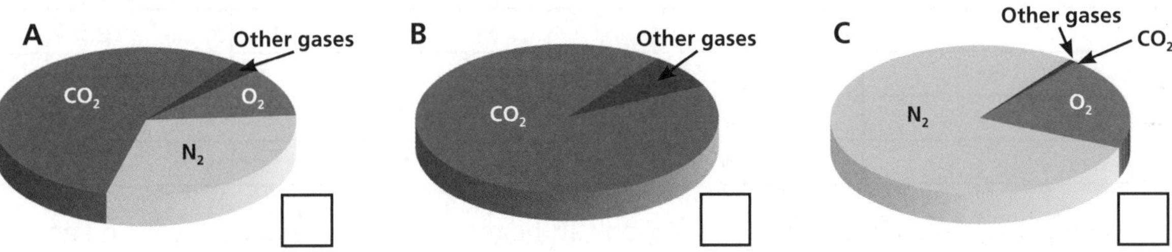

(c) Levels of carbon dioxide in the atmosphere are changing today. Explain how human activity is contributing to these changes. **[6]**

The quality of written communication will be assessed in your answer to this question.

2 Chlorine was first added to drinking water in England in the 1890s. Shortly after, the USA decided to do the same. The graph shows how the death rates from typhoid fever changed in the USA over a period of 60 years.

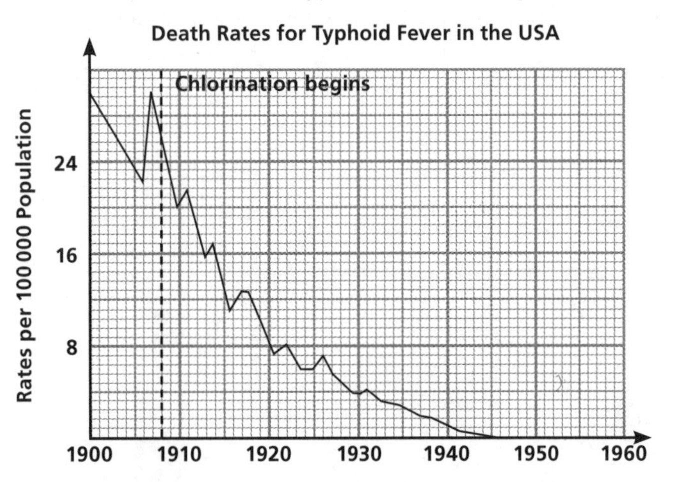

(a) In which year was chlorine first introduced into USA domestic water supplies? **[1]**

(b) Describe what happened to death rates for typhoid fever in the USA over the next 40 years. **[2]**

(c) How did the introduction of chlorine in water supplies produce these results? **[1]**

(d) Does the graph provide sufficient evidence to claim that there is a correlation between chlorinated water and typhoid fever? Explain your answer. **[4]**

3 Brine (sodium chloride solution) is an important industrial chemical. Passing an electric current through brine causes a chemical change to take place.

Which **three** products come from brine? Put ticks (✓) in the boxes next to the correct answers. **[3]**

Hydrogen gas	☐	Solid sodium chloride	☐	Chlorine gas	☐
Sodium hydroxide	☐	Hydrochloric acid	☐	Oxygen gas	☐

HT

4 A supermarket needs to produce some new carrier bags. It needs to decide whether to use traditional polythene, a biodegradable material or a UV-degradable material. It decides to carry out a strength test and then find out what the public think. Here are the results:

Test Number	Maximum Weight (N)		
	Polythene	Biodegradable	UV-degradable
1	26.50	19.99	23.48
2	26.53	20.03	23.45
3	26.49	20.01	23.49
4	26.70	19.99	23.46

(a) The supermarket manager takes a quick look at the results and then decides not to do any follow-up work on the biodegradable results. Why do you think he made this decision? **[1]**

(b) How could the manager best estimate the true value of the strength of each bag? **[2]**

(c) The results for the polythene tests contain an outlier. Identify the outlier. **[1]**

(d) Calculate a best estimate of the strength of the polythene bags. **[2]**

(e) Calculate a best estimate of the strength of the UV-degradable bags. **[2]**

(f) The manager asked the public what they thought of the different bags. Here are some of the responses:

Using the information from the public and the results of the tests, what do you think the manager should do? **[6]**

The quality of written communication will be assessed in your answer to this question.

Module C4 (Chemical Patterns)

Theories of atomic structure can be used to explain the properties and behaviour of elements. This module looks at:

- patterns in the properties of the elements
- how to explain the patterns in the properties of the elements
- the development of the periodic table
- the structure of the atom and electron configuration
- ionic bonding
- balanced equations
- how chemists explain the properties of compounds of Group 1 and Group 7 elements.

The Periodic Table

Elements are the 'building blocks' of all materials.

The atoms of each element have a different proton number. The elements are arranged in order of ascending **atomic** (or **proton**) **number**, which gives repeating patterns in the properties of elements.

The Development of the Periodic Table

Knowledge of chemical facts began to grow in the 18th and 19th centuries. As a result, scientists tried to find patterns in order to stop themselves being overwhelmed by the mass of information and to provide a basis for understanding the facts. Many of the early attempts of classification were dismissed by the scientific community as more information emerged.

Three of the most significant developments were made by Döbereiner, Newlands and Mendeleev.

Döbereiner (1829)
Döbereiner arranged elements in groups of three, known as triads. The elements in each triad had similar chemical properties, and the relative atomic mass of the middle one was about halfway between the other two.

For example:
Li – 7; Na – 23; K – 39
Cl – 35.5; Br – 80; I – 127

Atomic mass at that time was measured relative to the mass of a hydrogen atom. It is now measured relative to $\frac{1}{12}$ the mass of a carbon atom.

John Newlands (1864)
Newlands only knew of the existence of 63 elements; many were still undiscovered. He arranged the known elements in order of relative atomic mass and found similar properties amongst every eighth element in the series. This makes sense since the noble gases (Group 0) weren't discovered until 1894.

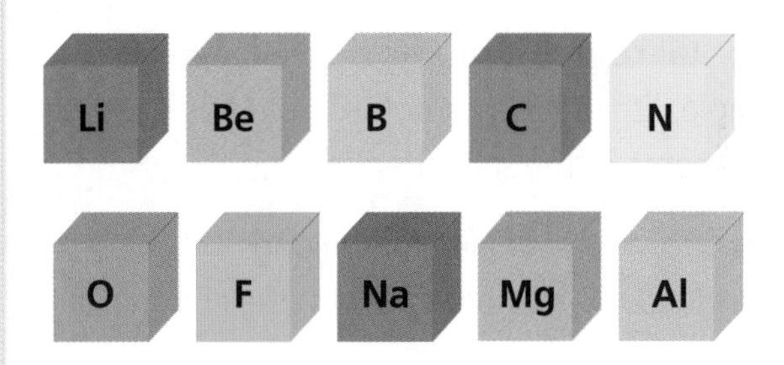

He had noticed some patterns, but the missing elements caused problems. Unfortunately his idea of putting elements into groups of eight had flaws and society greeted his idea with ridicule. The reasons for the flaws were later explained by the discovery of new elements, incorrect data and more complicated patterns after calcium.

Dimitri Mendeleev (1869)
Mendeleev realised that some elements had yet to be discovered, so he left gaps to accommodate their eventual discovery.

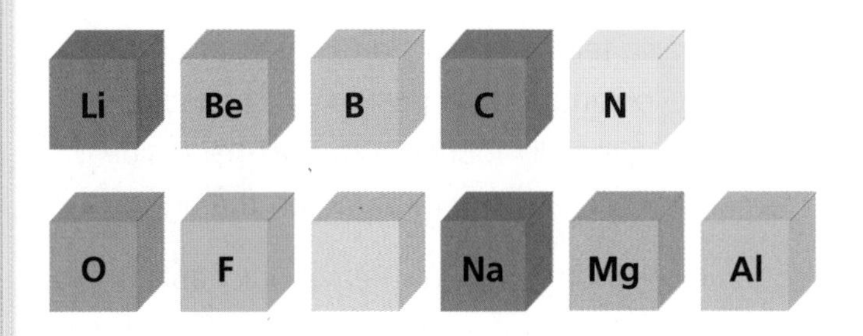

He used his periodic table to predict the existence of other elements. He also challenged some of the previous atomic mass data as being inaccurate.

Today's Periodic Table

Groups

A **vertical column** of elements is called a **group**. For example, lithium (Li), sodium (Na) and potassium (K) are all elements found in Group 1.

Elements in the same group have the same number of electrons in their outer shell (except helium). This number also coincides with the group number. For example, Group 1 elements have one electron in their outer shell, and Group 7 elements have seven electrons in their outer shell. Elements in the same group have similar properties.

Periods

A **horizontal row** of elements is called a **period**. For example, lithium (Li), carbon (C) and neon (Ne) are all elements in the same period.

The period to which an element belongs corresponds to the number of shells of electrons it has. For example, sodium (Na), aluminium (Al) and chlorine (Cl) all have three shells of electrons, so they are found in the third period.

The periodic table can be used as a reference table to obtain important information about the elements. For example:

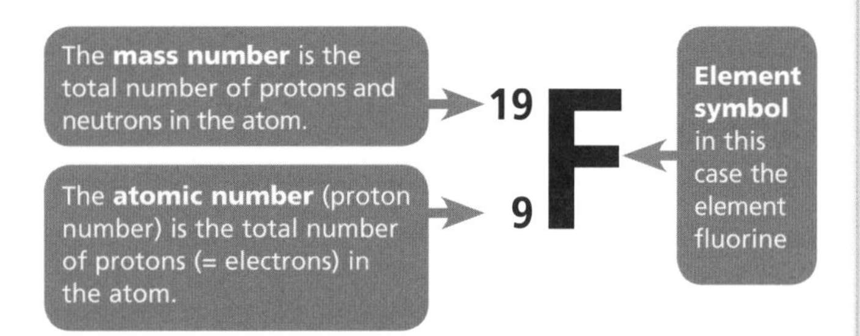

You can also tell if elements are metals or non-metals by looking at their position in the table. Metals are generally found on the left-hand side of the periodic table and non-metals on the right.

You will be given a copy of the periodic table in the exam. You can find a copy at the back of this book.

Atoms

Chemists use **atoms** to explain the properties of the elements. All substances are made up of atoms (very small particles). Each atom has a small central nucleus, made up of **protons** and **neutrons** (with the exception of hydrogen), which is surrounded by **electrons** arranged in **shells** (or **energy levels**).

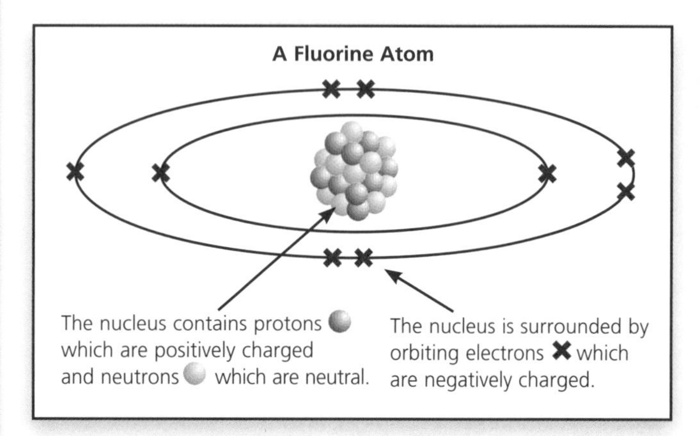

Atomic Particle	Relative Mass	Relative Charge
Proton	1	+1
Neutron	1	0
Electron ✖	0 (nearly)	-1

- An atom has the same number of protons as electrons, so the atom as a whole is neutral (i.e. it has no electrical charge).
- A proton has the same mass as a neutron.
- The mass of an electron is negligible, i.e. nearly nothing, when compared to a proton or neutron.
- A substance that contains only one sort of atom is called an element.
- All atoms of the same element have the same number of protons.
- Atoms of different elements have different numbers of protons.
- The elements are arranged in the periodic table in order of increasing atomic (proton) number.

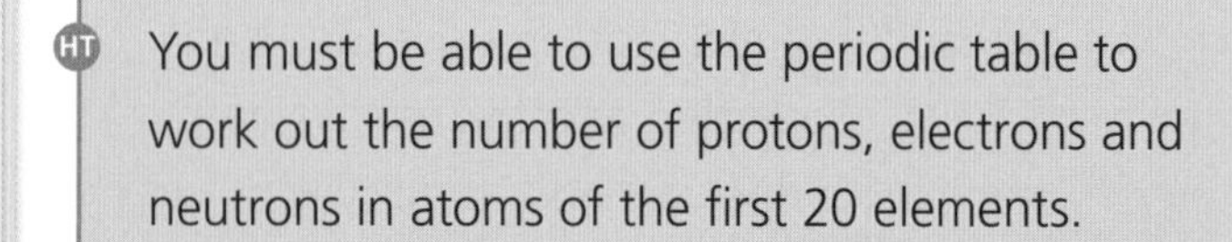

HT You must be able to use the periodic table to work out the number of protons, electrons and neutrons in atoms of the first 20 elements.

Spectroscopy

When some elements are heated, they emit distinctive coloured flames. Lithium, sodium, and potassium compounds can be recognised by the distinctive colours they produce in a flame test.

A piece of nichrome (a nickel-chromium alloy) wire is dipped in the compound and then put into a Bunsen flame to produce the following distinctive colours:

Lithium
Red

Sodium
Yellow

Potassium
Lilac

The light emitted from the flame of an element produces a characteristic line spectrum. Each line in the spectrum represents an energy change as excited electrons fall from high energy levels to lower energy levels. The study of spectra has helped chemists to discover new elements. The discovery of some elements depended on the development of new practical techniques like spectroscopy. White light shows a continuous spectrum, but the spectrum of a lithium flame is made up of a series of lines:

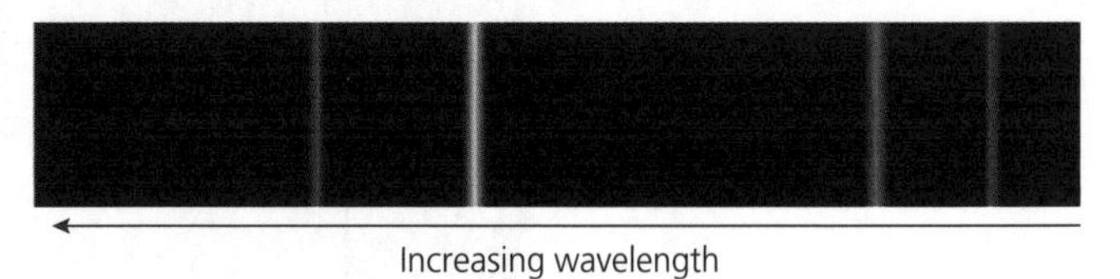

Electron Configuration

Electron configuration tells us how the electrons are arranged around the nucleus of an atom in shells (energy levels).

- The electrons in an atom occupy the lowest available shells (i.e. the shells closest to the nucleus).
- The first level (or shell) can only contain a maximum of two electrons.
- The second shell can hold a maximum of eight electrons.
- The third shell can hold up to 18 electrons (however, the last 10 of these electrons are only filled up after the first two in the fourth shell).
- The electron configuration is written as a series of numbers, e.g. oxygen is 2.6; aluminium is 2.8.3; and potassium is 2.8.8.1.

There is a connection between the number of outer electrons and the group. For example, an element with one electron in its outer shell is found in Group 1. You can also deduce the period to which an element belongs from its electron configuration. For example, oxygen is 2.6, which means it is found in the second period; potassium is 2.8.8.1, which means it is found in the fourth period, and so on (i.e. the number of energy levels occupied by electrons equals the number of the period).

HT The chemical properties of an element are determined by its electron arrangement.

Electron Configuration of the First 20 Elements

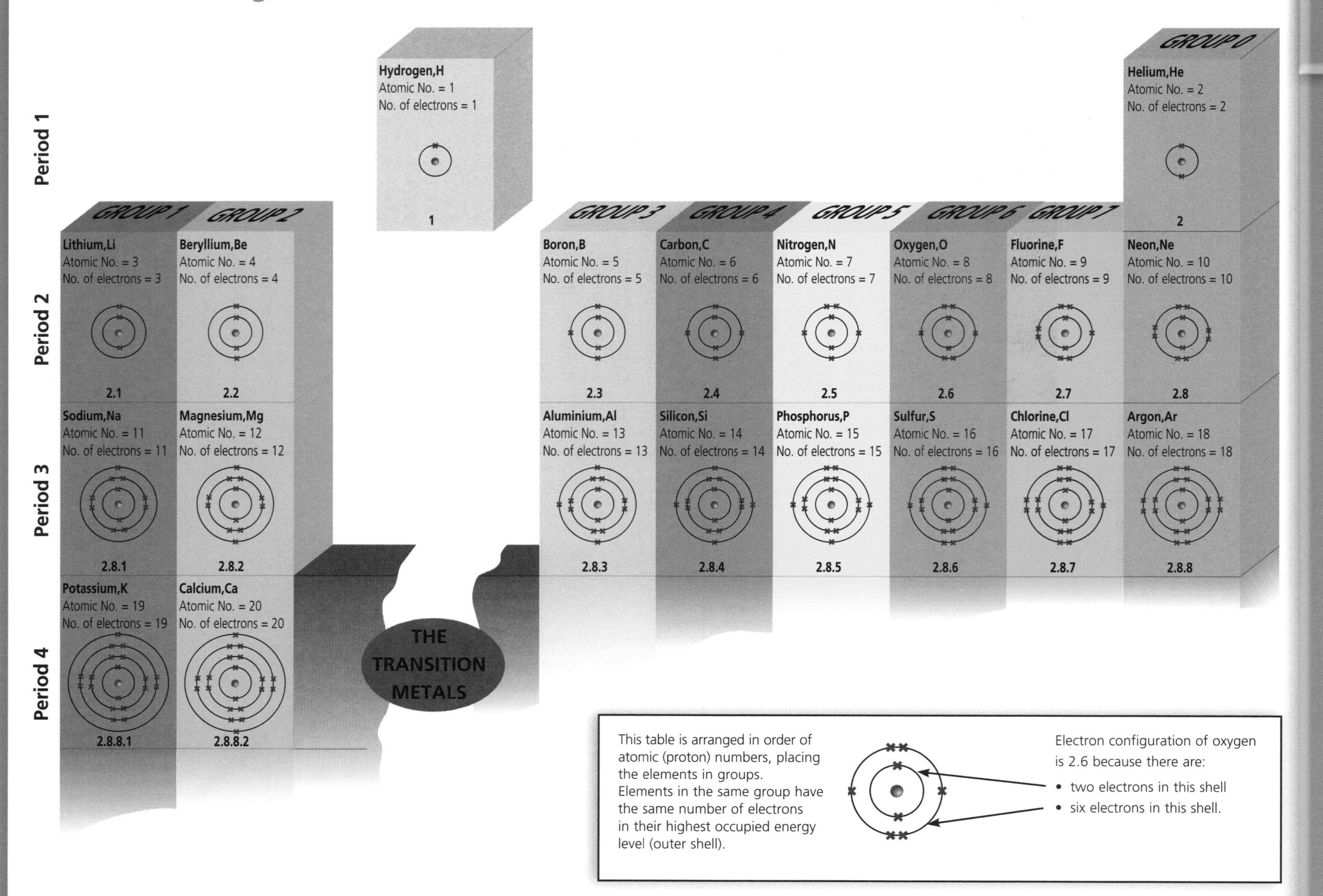

Hazard Symbols

Hazardous materials will have one of the following hazard symbols on their packaging.

Toxic	These substances can kill when swallowed, breathed in or absorbed through the skin.	
Oxidising	These substances provide oxygen, which allows other substances to burn more fiercely.	
Harmful	These substances are similar to toxic substances, but they are less dangerous.	
Highly Flammable	These substances will catch fire easily. They pose a serious fire risk.	
Corrosive	These substances attack living tissue, including eyes and skin, and can damage materials.	
Explosive	These substances will explode when they are set alight.	
Environmental Hazard	These substances may present a danger to the environment.	

Safety Precautions

Some common safety precautions are:

- wearing gloves and eye protection, and washing hands after handling chemicals
- using safety screens
- using small amounts and low concentrations of the chemicals
- working in a fume cupboard or ventilating the room
- not eating or drinking when working with chemicals
- not working near naked flames.

C4 Chemical Patterns

Group 1 – The Alkali Metals

There are six metals in Group 1. As we go down the group, the alkali metals become more reactive.

Alkali metals have low melting points. The melting and boiling points decrease as we go down the group.

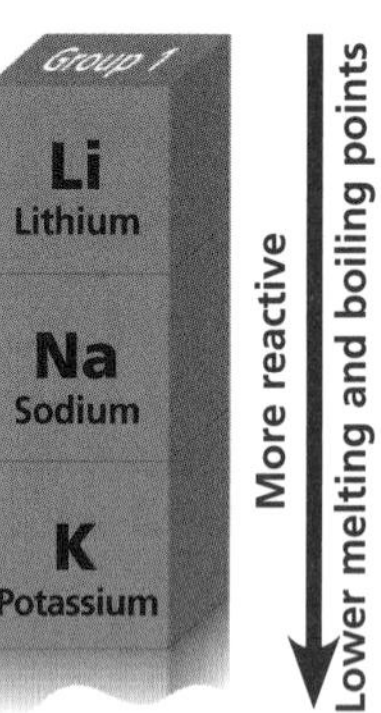

Physical Properties of the Alkali Metals

Element	Melting Point (°C)	Boiling Point (°C)	Density (g/cm^3)
Lithium, Li	180	1340	0.53
Sodium, Na	98	883	0.97
Potassium, K	64	760	0.86
Rubidium, Rb	39	688	1.53
Caesium, Cs	29	671	1.90

Alkali Metal Compounds

When alkali metals react they form compounds that are similar. The reaction becomes more vigorous as we go down the group because the metals are more reactive.

The table below shows the compounds formed when alkali metals react with water, oxygen and chlorine.

Element	+ Water	+ Oxygen	+ Chlorine
Lithium, Li	LiOH	Li_2O	LiCl
Sodium, Na	NaOH	Na_2O	NaCl
Potassium, K	KOH	K_2O	KCl

Reaction of Alkali Metals with Chlorine

Chlorine reacts vigorously with alkali metals to form colourless crystalline salts called metal chlorides, for example:

Lithium + Chlorine ⟶ Lithium chloride
$2Li_{(s)} + Cl_{2(g)} \longrightarrow 2LiCl_{(s)}$

You must be able to interpret symbol equations like the one above.

HT You must be able to write balanced equations for the reactions of the alkali metals and halogens.

Reaction of Alkali Metals with Oxygen

The alkali metals are stored under oil because they react ve vigorously with oxygen and water. When freshly cut, they are shiny. However, they quickly tarnish in moist air, go dul and become covered in a layer of metal oxide. For example

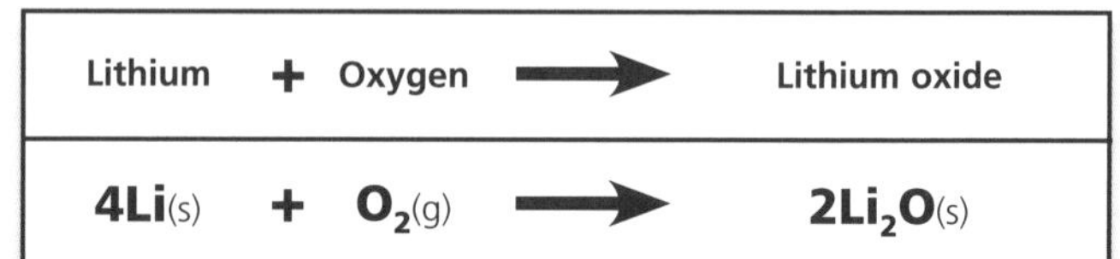

Reaction of Alkali Metals with Water

Lithium, sodium and potassium float on top of cold water (due to their low density). The heat from the reaction is great enough to turn sodium and potassium into liquids. Lithium reacts gently, sodium more aggressively and potassium so aggressively it catches fire.

When alkali metals react with water, a metal hydroxide and hydrogen gas are formed. The metal hydroxide dissolves in water to form an alkaline solution, for example:

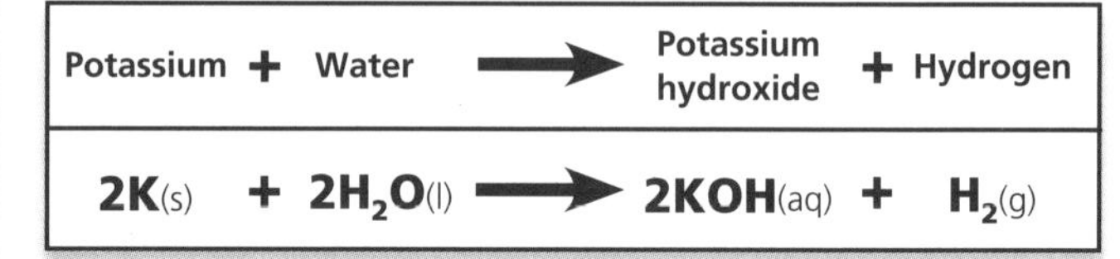

Follow these steps to check the pH level of the solution formed when an alkali metal is added to water.

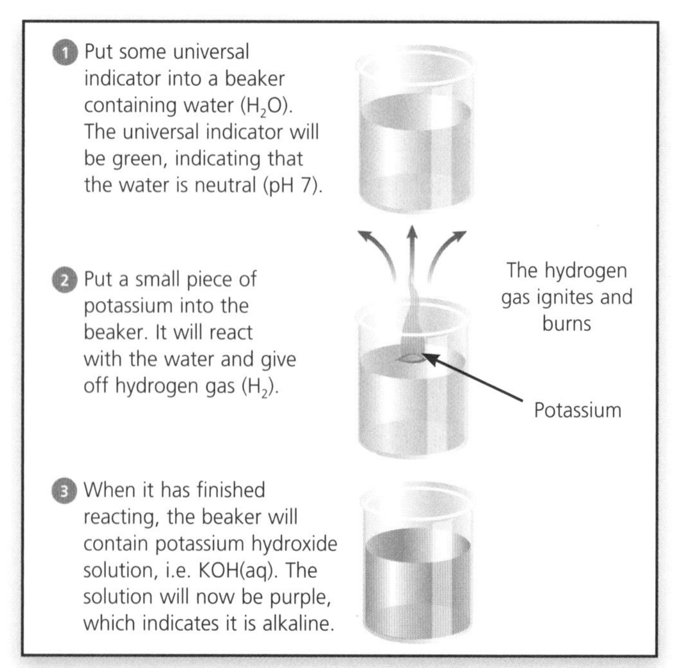

Since all the reactions of alkali metals are similar, a general equation or formula is sometimes used. In each case M refers to the alkali metal.

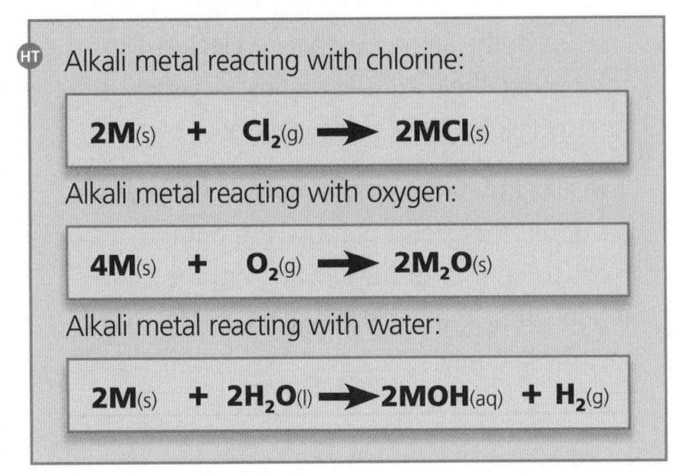

The general formula of the chlorides of the alkali metals is MCl, and the general formula of the hydroxides of the alkali metals is MOH. You need to be able to interpret symbols like these, as well as symbol equations.

The following hazard symbols are found on alkali metals and their compounds:

Chemical	Hazard Symbol
Lithium	
Lithium chloride	
Sodium	
Sodium hydroxide	
Potassium	

Therefore when working with Group 1 metals, the following precautions should be taken:

- Use small amounts of very dilute concentrations.
- Wear safety glasses and use safety screens.
- Watch teacher demonstrations carefully.
- Avoid working near naked flames.

Group 7 – The Halogens

There are five non-metals in Group 7.

At room temperature and room pressure, chlorine is a green gas, bromine is an orange liquid and iodine is a purple / dark grey solid. When heated, bromine forms a brown gas and iodine a pale purple gas.

Chlorine is used to sterilise water and to make pesticides and plastics.

Iodine is used as an antiseptic to sterilise wounds.

All halogens consist of **diatomic molecules** (i.e. they only exist in pairs of atoms), e.g. Cl_2, Br_2 and I_2. They can be used to bleach dyes and kill bacteria in water.

Physical Properties of Halogens

The physical properties of the halogens alter as we go down the group. The table below shows their melting points, boiling points and densities:

Element	Melting Point (°C)	Boiling Point (°C)	Density (g/cm^3)
Fluorine, F	-220	-188	0.0016
Chlorine, Cl	-101	-34	0.003
Bromine, Br	-7	59	3.12
Iodine, I	114	184	4.95
Astatine, At	302	337	7 (estimated)

- The melting points and boiling points of the halogens increase as we go down the group.
- The densities increase as we go down the group. (However, due to the unstable nature of astatine, its density is estimated.)

Hazards of the Halogens

The halogens carry the following hazard symbols:

Element	Hazard Symbol
Chlorine, Cl	
Bromine, Br	
Iodine, I	

Therefore when working with halogens, the following precautions should be taken:

- Wear safety glasses.
- Work in a fume cupboard.
- Make sure the room is well ventilated.
- Use small amounts of very dilute concentrations.
- Avoid working near naked flames.
- Watch teacher demonstrations carefully.

Displacement Reactions of Halogens

A more reactive halogen will displace a less reactive halogen from an aqueous solution of its salt. Therefore, chlorine will displace both bromine and iodine, while bromine will displace iodine.

Potassium iodide	+	Chlorine	→	Potassium chloride	+	Iodine
$2KI(aq)$	+	$Cl_2(aq)$	→	$2KCl(aq)$	+	$I_2(aq)$

Halogen Compounds

When halogens react, they form compounds that are similar. The reactivity decreases as we go down the group. For example, this is seen clearly when halogens react with the alkali metals or iron. The reaction between chlorine and iron is more vigorous than that between iodine and iron.

The table below shows the compounds that are formed when halogens react with Group 1 metals.

	Chlorine	Bromine	Iodine
Lithium	LiCl	HT LiBr	LiI
Sodium	NaCl	HT NaBr	NaI
Potassium	KCl	HT KBr	KI

HT Trends in Group 1

Alkali metals have similar properties because they have the same number of electrons in their outer shell, i.e. the highest occupied energy level contains one electron.

The alkali metals become more reactive as we go down the group because the outer electron shell is further away from the influence of the nucleus and so an electron is lost more easily.

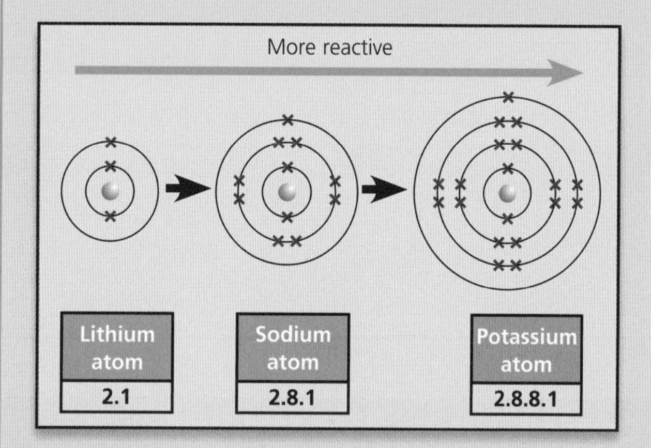

Trends in Group 7

The halogens have similar properties because they have the same number of electrons in their outer shell, i.e. the highest occupied energy level contains seven electrons.

The halogens become less reactive down the group because the outer electron shell is further away from the influence of the nucleus, so an electron is gained less easily.

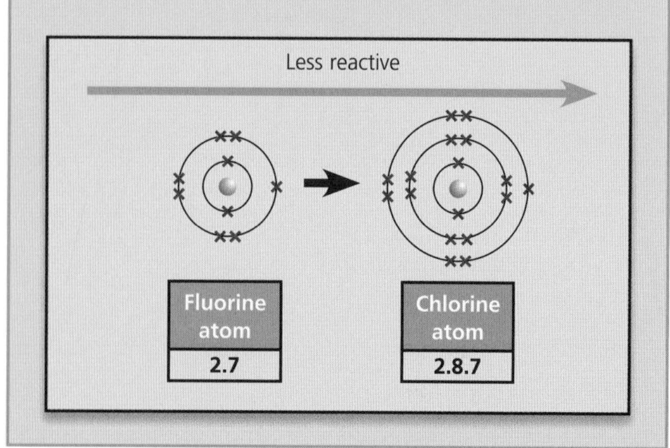

HT Balanced Equations

The total mass of the products of a chemical reaction is always equal to the total mass of the reactants.

This is because the products of a chemical reaction are made up from the atoms of the reactants – no atoms are lost or made. So, chemical symbol equations must always be balanced: there must be the same number of atoms of each element on the reactant side of the equation as there is on the product side.

Writing Balanced Equations

Follow these steps to write a balanced equation:

1. Write a word equation for the chemical reaction.
2. Substitute in formulae for the elements or compounds involved.
3. Balance the equation by adding numbers in front of the reactants and / or products.
4. Write a balanced symbol equation.

Example 1

	Reactants			→	Products
Write a word equation	Magnesium	+	Oxygen	→	Magnesium oxide
Substitute in formulae	$Mg(s)$	+	$O_2(g)$	→	$MgO(s)$
Balance the equation	Mg	+	O O	→	Mg O
	Mg	+	O O	→	Mg O Mg O
	Mg Mg	+	O O	→	Mg O Mg O
Write a balanced symbol equation	$2Mg(s)$	+	$O_2(g)$	→	$2MgO(s)$

- There are two **O**s on the reactant side, but only one **O** on the product side. We need to add another **MgO** to the product side to balance the **O**s.
- We now need to add another **Mg** on the reactant side to balance the **Mg**s.
- There are two **Mg** atoms and two **O** atoms on each side – **it is balanced**.

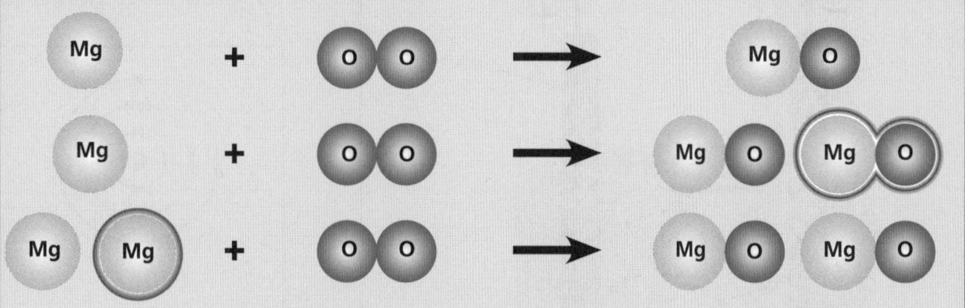

Example 2

	Reactants			→	Products		
Word equation	Sodium	+	Water	→	Sodium hydroxide	+	Hydrogen
Substitute in formulae	$Na(s)$	+	$H_2O(l)$	→	$NaOH(aq)$	+	$H_2(g)$
Balanced symbol equation	$2Na(s)$	+	$2H_2O(l)$	→	$2NaOH(aq)$	+	$H_2(g)$
	Na Na	+	H O H, H O H	→	H O Na, Na O H	+	H H
This means that	Two atoms of sodium which are solid	and	Two molecules of water which are liquid	produce	Two sodium hydroxides in aqueous solution	and	One molecule of hydrogen which is a gas

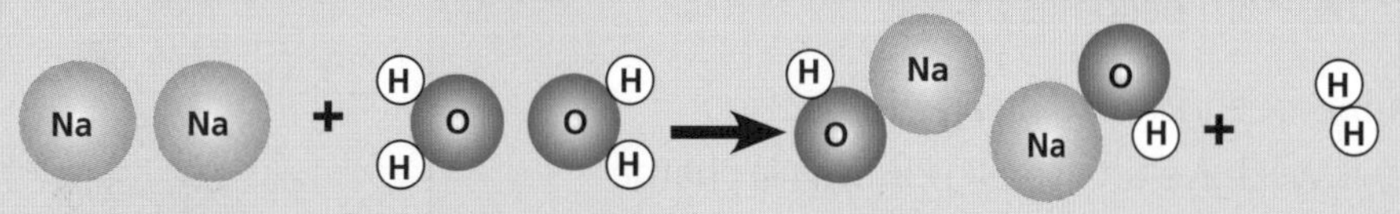

(s), (l), (aq) and (g) are the state symbols

The Properties of Compounds

Chemists use their observations to develop theories to explain the properties of different compounds. For example, experiments show that molten compounds of metals with non-metals, such as lithium chloride, conduct electricity.

It can, therefore, be concluded that there must be charged particles in molten compounds. These particles are known as **ions**.

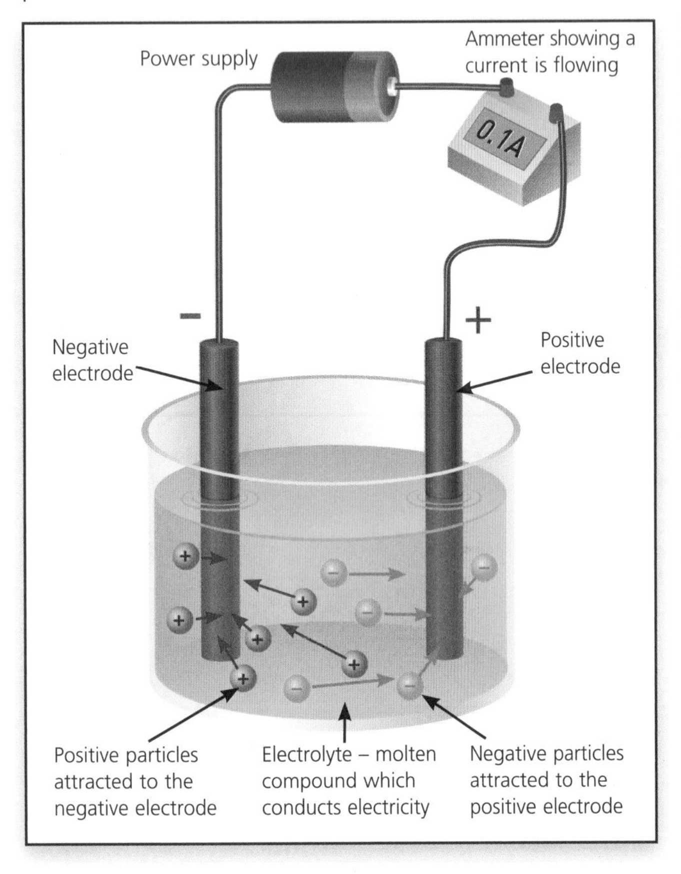

Ions

If an atom loses or gains one or more electrons, it will carry an overall charge because the proton and electron numbers are no longer equal. When this happens, the atom becomes an ion.

If the ion has been formed by an atom losing electrons, it will have an overall positive (+) charge because it now has more protons than electrons. It is called a **cation**, e.g. Na^+. If the ion has been formed by an atom gaining electrons, it will have an overall negative (–) charge because it now has more electrons than protons. It is called an **anion**, e.g. Cl^-.

The Ionic Bond

An ionic bond occurs between a metal and a non-metal and involves the transfer of electrons from one atom to another to form electrically charged ions.

Each electrically charged ion has a complete outer energy level or shell, i.e. the first shell has two electrons and each outer shell has eight electrons. Compounds of Group 1 metals and Group 7 elements are **ionic compounds**.

Example 1

Sodium and chlorine bond ionically to form sodium chloride, NaCl. The sodium (Na) atom has one electron in its outer shell which is transferred to the chlorine (Cl) atom, so they both have eight electrons in their outer shell. The atoms become ions, Na^+ and Cl^-, and the compound formed is sodium chloride, NaCl.

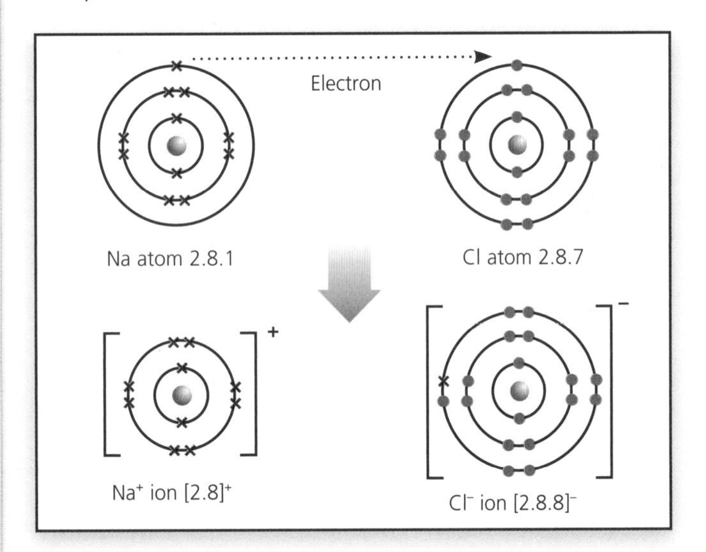

The positive ion and the negative ion are then electrostatically attracted to each other to form a giant crystal lattice.

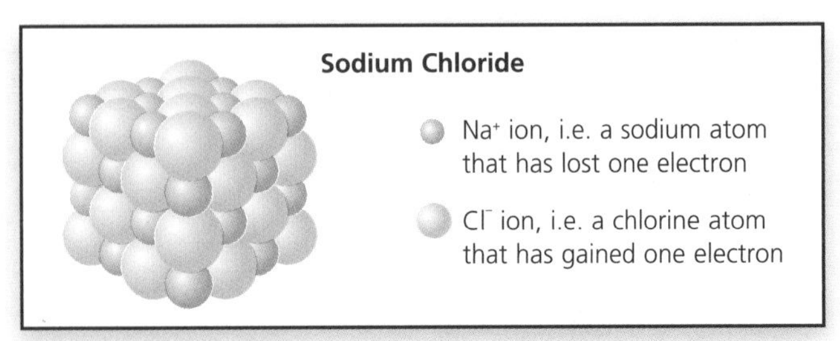

Sodium chloride has a high melting point and dissolves in water. It conducts electricity when it is in solution or is molten, but not when it is a solid.

Example 2

Sodium and oxygen bond ionically to form sodium oxide, Na_2O. Each sodium (Na) atom has one electron in its outer shell. An oxygen (O) atom wants two electrons, therefore two Na atoms are needed. The atoms become ions (Na^+, Na^+ and O^{2-}) and the compound formed is sodium oxide, Na_2O.

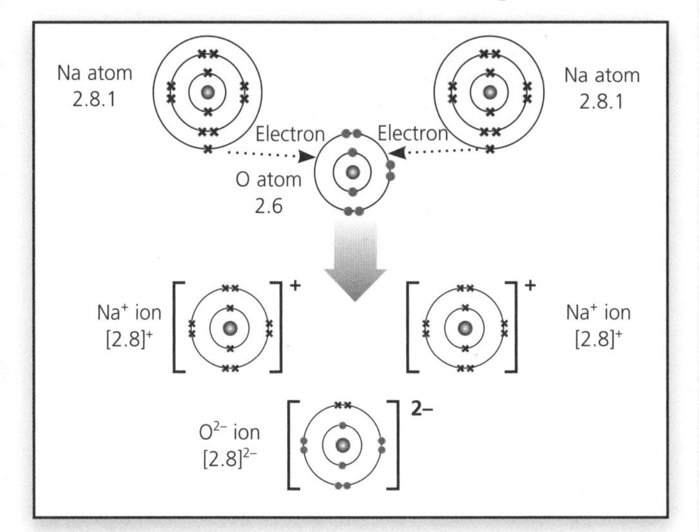

Example 3

Magnesium and oxygen bond ionically to form magnesium oxide, MgO. The magnesium (Mg) atom has two electrons in its outer shell. The oxygen atom only has six electrons in its outer shell, so the two electrons from the Mg atom are transferred to the O atom, so they both have eight electrons in their outer shell. The atoms become ions (Mg^{2+} and O^{2-}) and the compound formed is magnesium oxide, MgO.

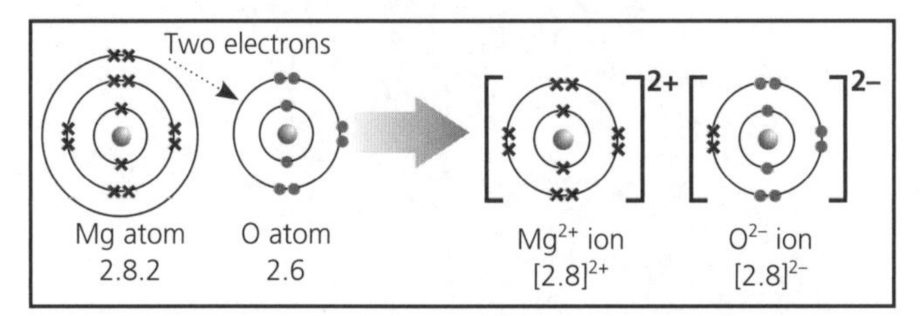

Many ionic compounds have properties similar to those of sodium chloride. They form crystals because the ions are arranged into a regular lattice.

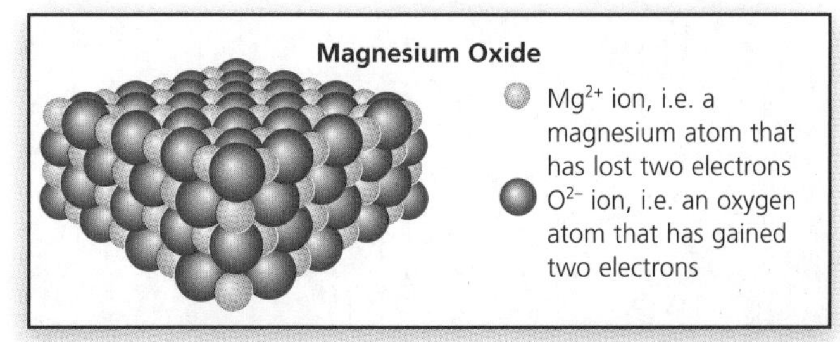

There is a strong force of attraction between the ions, which takes a lot of energy to break, so ionic compounds have high melting and boiling points. When the compound is molten or dissolved, the forces of attraction are weakened so much that the charged ions are free to move. This means they can conduct electricity.

HT Deducing the Formula of an Ionic Compound

If you know the charge on both ions, you can work out the formula of an ionic compound. If you know the formula and the charge on one of the ions, you can work out the charge on the other ion. This is because all ionic compounds are neutral substances where the charge on the positive ion(s) is equal to the charge on the negative ion(s).

Positive Ions	Negative Ions: 1– e.g. Cl^-, OH^-	Negative Ions: 2– e.g. SO_4^{2-}, O^{2-}
1+ e.g. K^+, Na^+	KCl: 1+, 1–	K_2SO_4: 2 × 1+ = 2+, 2–
	NaOH: 1+, 1–	Na_2O: 2 × 1+ = 2+, 2–
2+ e.g. Mg^{2+}, Cu^{2+}	$MgCl_2$: 2+, 2 × 1– = 2–	$MgSO_4$: 2+, 2–
	$Cu(OH)_2$: 2+, 2 × 1– = 2–	CuO: 2+, 2–
3+ e.g. Al^{3+}, Fe^{3+}	$AlCl_3$: 3+, 3 × 1– = 3–	$Al_2(SO_4)_3$: 2 × 3+ = 6+, 3 × 2– = 6–
	$Fe(OH)_3$: 3+, 3 × 1– = 3–	Fe_2O_3: 2 × 3+ = 6+, 3 × 2– = 6–

C5 Chemicals of the Natural Environment

Module C5 (Chemicals of the Natural Environment)

We can get a better understanding of the impact human activity can have on the natural environment by knowing more about the chemicals that make up our planet. This module looks at:

- the structure and properties of the chemicals found in the atmosphere, hydrosphere and lithosphere
- how to extract useful metals from minerals
- the structure and properties of metals.

The Earth's Resources

1. **Atmosphere** – a layer of gases surrounding the Earth. It is made up of the elements nitrogen (N), oxygen (O), traces of argon (Ar) and some compounds, e.g. carbon dioxide (CO_2) and water vapour (H_2O).

2. **Hydrosphere** – all the water on the Earth, including oceans, seas, rivers, lakes and underground reserves. The water contains dissolved compounds.

3. **Lithosphere** – the rigid outer layer of the Earth made up of the crust and the part of the mantle just below it. It is a mixture of minerals, such as silicon dioxide (SiO_2). Abundant elements in the lithosphere include silicon (Si), oxygen (O) and aluminium (Al).

Chemicals of the Atmosphere

Chemical (% composition)	2D Molecular Diagram	3D Molecular Diagram	Boiling Point (°C)	Melting Point (°C)
Oxygen, O_2 (21%)	O=O		-182.9	-218.3
Nitrogen, N_2 (78%)	N≡N		-195.8	-210.1
Carbon dioxide, CO_2 (varies, approx 0.035%)	O=C=O		-78.0	Sublimes (no liquid state)
Water vapour, H_2O (varies from 0–4%)	H–O–H		100.0	0
Argon, Ar (0.93%)	Ar		-185.8	-189.3

The chemicals that make up the atmosphere consist of non-metal elements and molecular compounds made up from non-metal elements.

From the information in the table above we can deduce that the molecules (with the exception of water) that make up the atmosphere are gases at 20°C because they have very low boiling points, i.e. they boil below 20°C. This can be explained by looking at the structure of the molecules. Molecular compounds have strong covalent bonds between the atoms that make up the compound but only weak forces of attraction between the small molecules.

So, for example, gases consist of small molecules with weak forces of attraction between the molecules. Only small amounts of energy are needed to break these forces, which allows the molecules to move freely through the air.

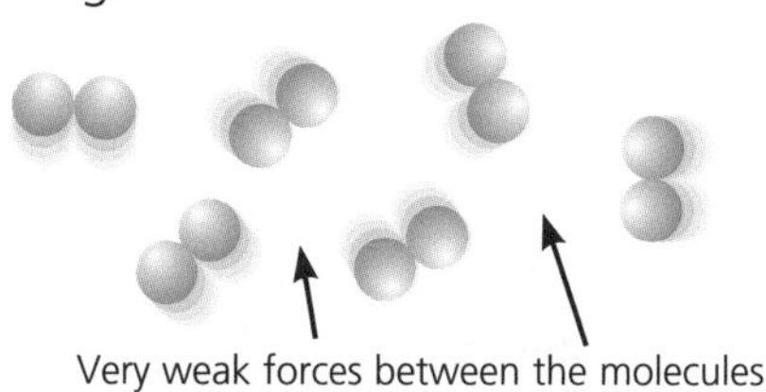

The atoms within molecules (e.g. hydrogen) are connected by strong covalent bonds.

HT In covalent bonds, electrons are shared between the nuclei of two atoms. This causes a strong, electrostatic attraction between the nuclei and shared electrons.

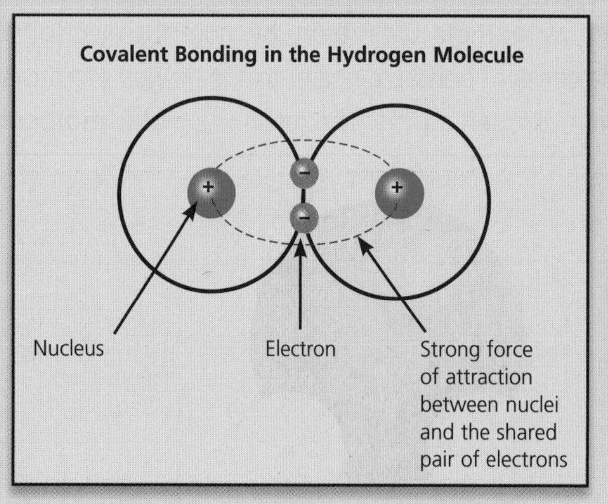

Unlike ionic compounds, pure molecular compounds do not conduct electricity because their molecules are not charged.

Chemicals of the Hydrosphere

Seawater in the hydrosphere is 'salty' because it contains dissolved ionic compounds. Examples of dissolved ionic compounds are:

- sodium chloride, NaCl
- magnesium chloride, $MgCl_2$
- magnesium sulfate, $MgSO_4$
- sodium sulfate, Na_2SO_4
- potassium chloride, KCl
- potassium bromide, KBr.

Examples of solid ionic compounds and how their physical properties relate to their giant structure were given in Module C4.

HT When given a table of charges, you must be able to work out the formulae for the following ionic compounds: sodium chloride, magnesium chloride, sodium sulfate, magnesium sulfate, potassium chloride and potassium bromide.

The Water Molecule

Water has some unexpected properties. For example, the table on the previous page shows that the boiling point of water is 100°C. This is a much higher boiling point than for the other molecules listed. Water is also a good **solvent** for salts. The properties can be explained by its structure.

The water molecule is bent, because the electrons in the covalent bond are nearer to the oxygen atom than the hydrogen atoms. The result is a **polar molecule**.

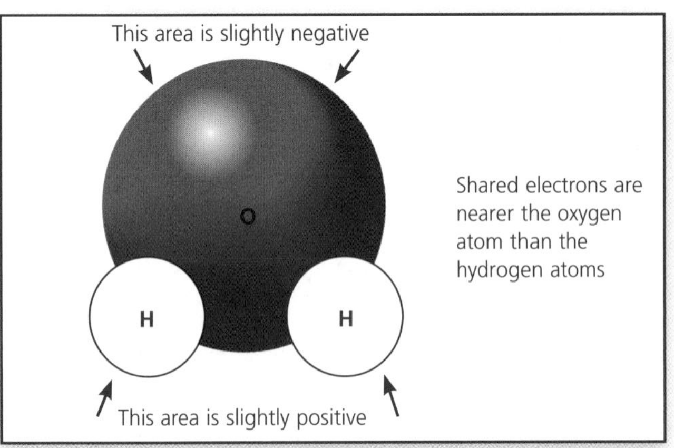

The small charges on the atoms mean that the forces between the molecules are slightly stronger than in other covalent molecules. Therefore, more energy is needed to separate them.

The small charges also help water to **dissolve** ionic compounds as the water molecules attract the charges on the ions. The ions can then move freely through the liquid.

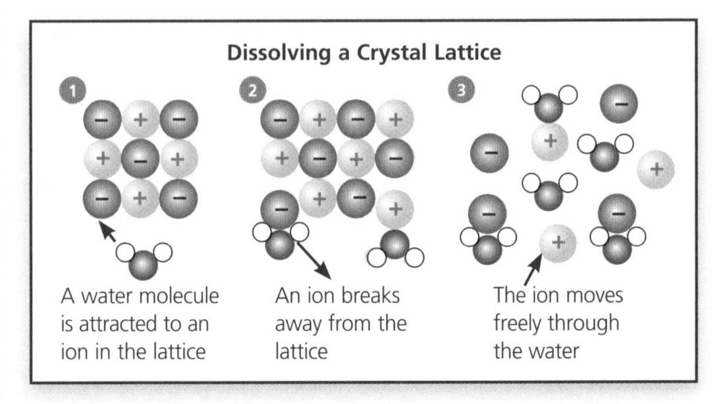

Identification of Ions

Ions in ionic compounds can be detected and identified because they have distinct properties and they form chemicals with distinct properties. For example, an insoluble compound may **precipitate** on mixing two solutions of ionic compounds. This technique is often used to identify metal ions. In the example below, a white precipitate of calcium hydroxide is formed (as well as sodium chloride solution). We can see how the precipitate is formed by considering the ions involved:

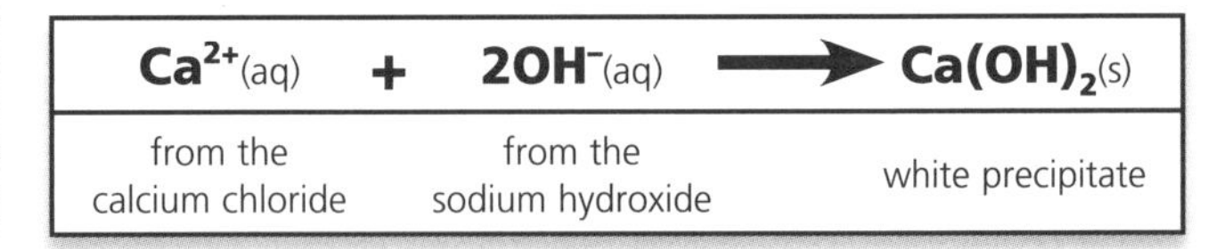

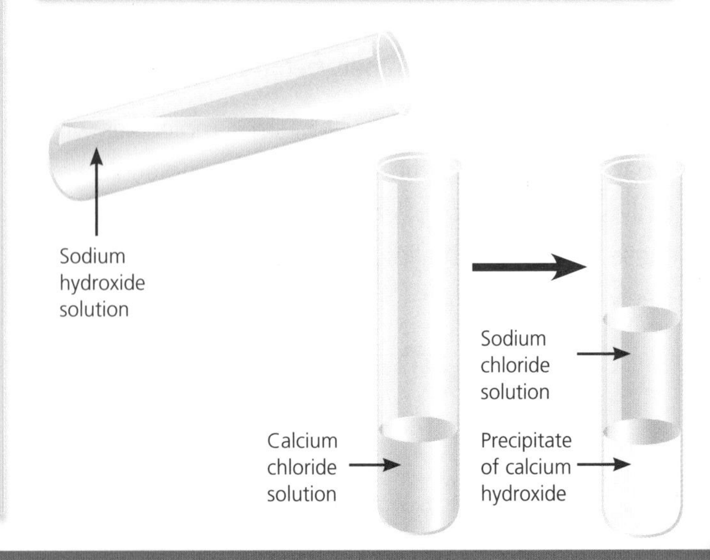

The table below shows the colour of the precipitate formed when certain metal ions are mixed with alkali.

Metal Ion + Sodium Hydroxide		Precipitate Formed	Colour of Precipitate
Aluminium	$Al^{3+}(aq)$	Aluminium hydroxide	White
Calcium	$Ca^{2+}(aq)$	Calcium hydroxide	White
Magnesium	$Mg^{2+}(aq)$	Magnesium hydroxide	White
Copper	$Cu^{2+}(aq)$	Copper hydroxide	Blue
Iron(II)	$Fe^{2+}(aq)$	Iron(II) hydroxide	Green
Iron(III)	$Fe^{3+}(aq)$	Iron(III) hydroxide	Brown

HT You must be able to write ionic equations for all the precipitate reactions and other reactions given in this module, when provided with appropriate information.

In the oceans, dissolved calcium ions and carbonate ions combine to form a precipitate of calcium carbonate (limestone), which falls to the ocean floor.

Calcium ion + Carbonate ion ⟶ Calcium carbonate

HT $Ca^{2+}(aq) + CO_3^{2-}(aq) \longrightarrow CaCO_3(s)$

In order to identify a negative ion, a range of different tests can be carried out. These involve adding a reagent to the unknown sample, which reacts with the ions to form an insoluble salt.

Sulfate Ions

To identify the presence of a sulfate ion, add barium chloride solution and dilute hydrochloric acid to the suspected sulfate solution.

A white precipitate of barium sulfate will be produced if a sulfate is present.

Barium ion + Sulfate ion ⟶ Barium sulfate

HT $Ba^{2+}(aq) + SO_4^{2-}(aq) \longrightarrow BaSO_4(s)$

White precipitate

Chloride, Bromide and Iodide Ions

To identify the presence of a chloride, bromide or iodide ion, add silver nitrate solution and nitric acid to the suspected halide solution. A white precipitate will form if silver chloride is present, a cream precipitate for silver bromide, and a yellow precipitate for silver iodide.

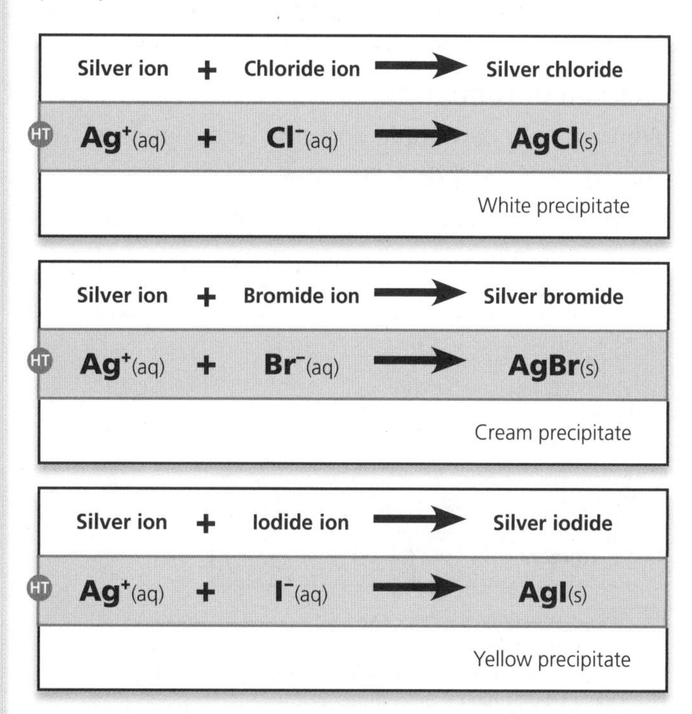

HT **Solubility of Ionic Compounds**

In general, most ionic compounds are soluble in water, but as you have seen there are some exceptions. Given information about solubility, you must be able to predict which chemicals are likely to precipitate out when mixing solutions of ionic compounds.

Testing for Carbonates

Testing with Acids

Carbonates react with dilute acids to form carbon dioxide gas (and a salt and water). For example, if we add calcium carbonate to dilute hydrochloric acid then the carbonate will 'fizz' as it reacts with the acid, giving off carbon dioxide.

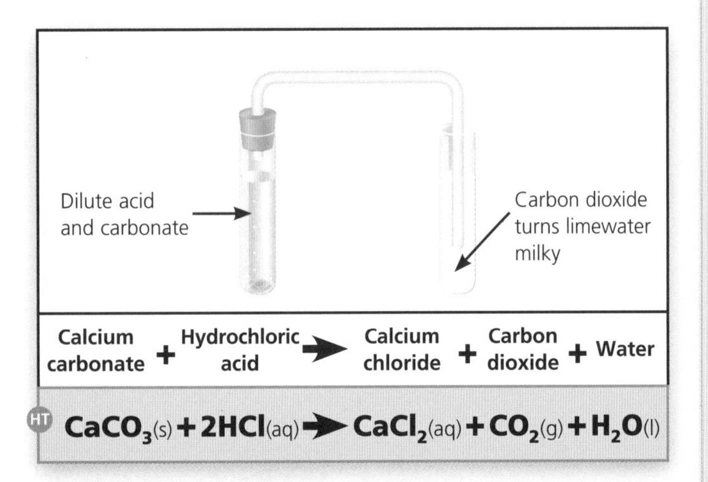

Thermal Decomposition

When copper carbonate and zinc carbonate are heated, a thermal decomposition reaction takes place. This results in a distinctive colour change, which enables the two compounds to be identified.

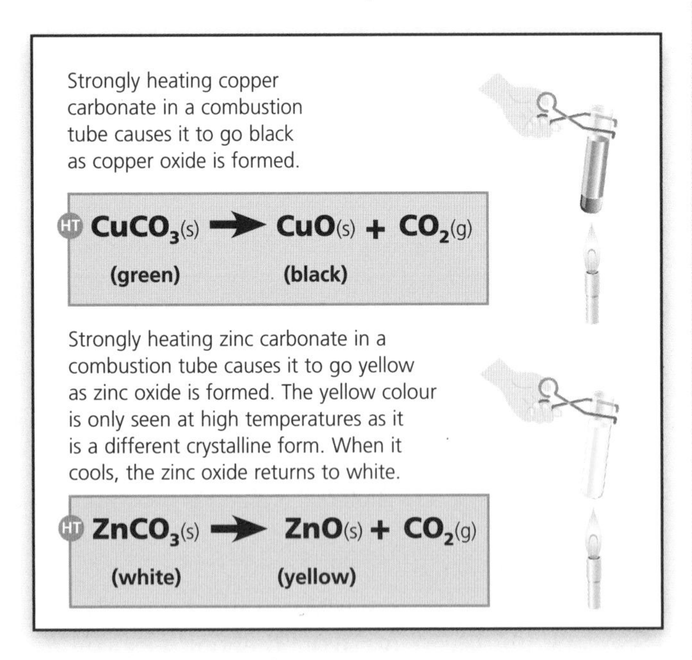

You must be able to interpret the information given on pages 42–43 to work out which ions are present in an unknown sample when you are given a test with the result.

Chemicals of the Lithosphere

Element	Abundance in Lithosphere (ppm)
Oxygen, O	455 000
Silicon, Si	272 000
Aluminium, Al	83 000
Iron, Fe	62 000
Calcium, Ca	46 600
Magnesium, Mg	27 640
Sodium, Na	22 700
Potassium, K	18 400
Titanium, Ti	6320
Hydrogen, H	1520

N.B. You may be asked to interpret data like this in your exam.

The table shows that the three most abundant elements in the Earth's crust are oxygen, silicon and aluminium. Much of the silicon and oxygen is present as the compound silicon dioxide (SiO_2).

Silicon Dioxide

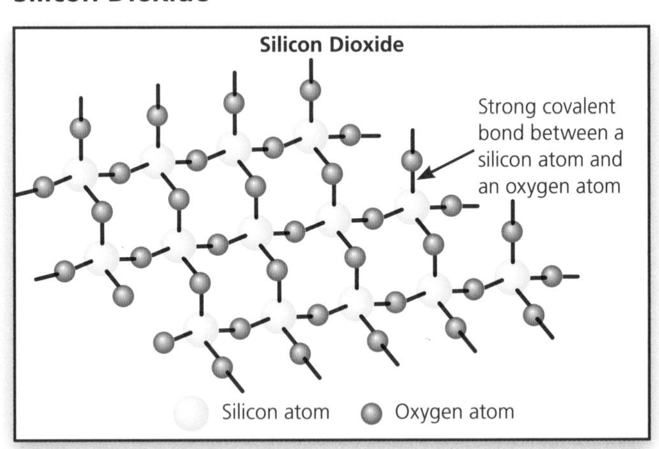

Silicon dioxide forms a **giant covalent structure**, where each silicon atom is covalently bonded to four oxygen atoms. Each oxygen atom is bonded to two silicon atoms. The result is a very strong, rigid three-dimensional structure, which is very difficult to break down. Silicon dioxide does not conduct electricity because there are no ions or free electrons in the structure. It does not dissolve in water because there are no charges to attract the water molecules. It also has high melting and boiling points.

Silicon dioxide exists in different forms, such as **quartz** in granite, and it is the main constituent of sandstone. Amethyst is a form of quartz that is used as a **gemstone**. It is cut, polished and used in jewellery. The violet colour comes from traces of manganese oxide and iron oxides found in the quartz. Some gemstones are very valuable because of their rarity, hardness and shiny appearance.

Carbon

Carbon is another example of a mineral that forms a giant covalent structure. Two forms of carbon are:

- diamond
- graphite.

Diamond

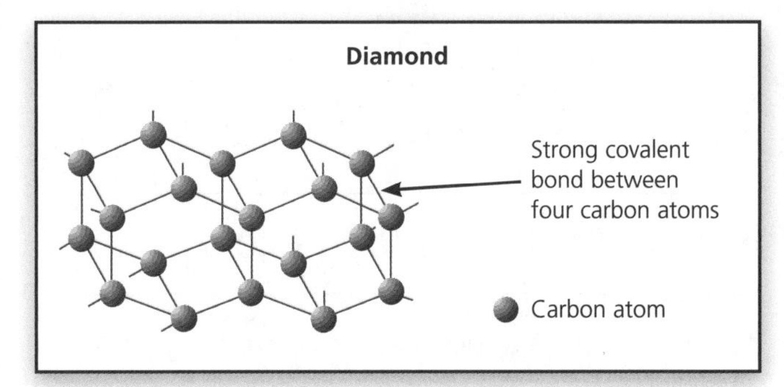

Diamond has a large number of covalent bonds, which means it has very high melting and boiling points. Each carbon atom is covalently bonded to four other carbon atoms, resulting in a very strong, rigid, three-dimensional structure that is difficult to break down. Diamond is insoluble because there are no charges to attract water molecules. It does not conduct electricity because there are no ions or free electrons in the structure.

Graphite

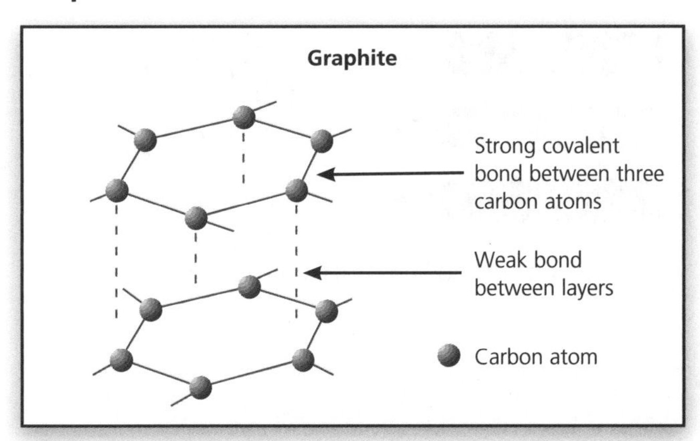

Graphite is a form of carbon that has a giant covalent structure. Each carbon atom is covalently bonded to three other carbon atoms in a layered structure. The layers can slide past each other, making it soft and slippery.

Graphite is also insoluble and has high melting and boiling points, as the carbon atoms in each layer are held together by strong covalent bonds. Graphite can conduct electricity because the electrons forming the weak bonds between the layers are free to move throughout the whole structure.

Using Chemicals of the Lithosphere

By understanding giant covalent structures, we can explain why we use different materials for certain jobs. The table gives some examples:

Element / Compound	Property	Use	Explanation
Carbon – diamond	Very hard	Drill tips	A lot of energy is needed to break the strong covalent bonds between the atoms.
Carbon – graphite	Soft	Pencils	Layers easily removed and stick to paper.
Silicon dioxide	High melting point (1610°C)	Furnace linings	A lot of energy is needed to break the strong covalent bonds between the atoms.
Silica glass	Does not conduct electricity	Insulator in electrical devices	No free electrons or ions to carry electrical charge.

HT Balancing Equations

All chemical reactions follow the same simple rule: the mass of the reactants is equal to the mass of the products. This means there must be the same number of atoms on both sides of the equation.

You do not have to do anything if the equation is already balanced. If the equation needs balancing, follow this method:

1. Write a number in front of one or more of the formulae. This increases the number of all of the atoms in that formula.
2. Include the state symbols: (s) = solid, (l) = liquid, (g) = gas and (aq) = dissolved in water (aqueous solution).

Example
Balance the reaction between calcium carbonate and hydrochloric acid.

Calcium carbonate	+	Hydrochloric acid	➔	Calcium chloride	+	Carbon dioxide	+	Water
$CaCO_3$	+	HCl	➔	$CaCl_2$	+	CO_2	+	H_2O

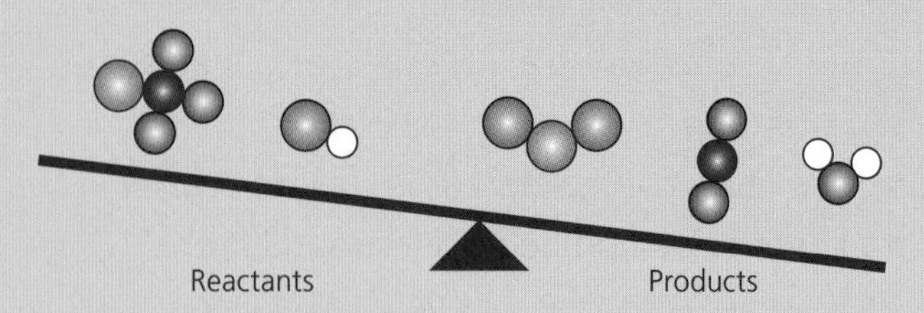

There are more chlorine atoms and hydrogen atoms on the products side than on the reactants side, so balance chlorine by doubling the amount of hydrochloric acid.

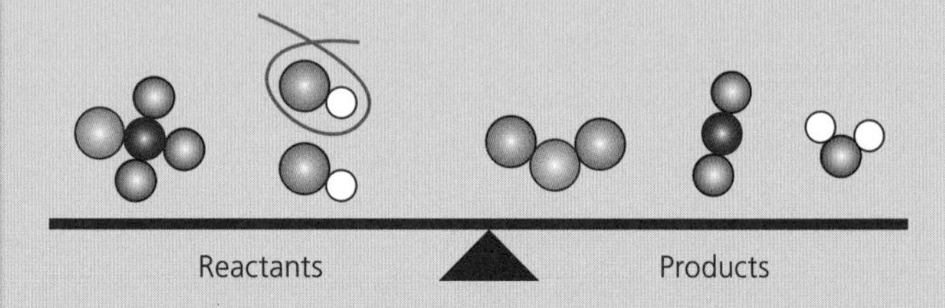

The amount of chlorine and hydrogen on both sides is now equal. This gives you a balanced equation.

$$CaCO_{3(s)} + 2HCl_{(aq)} \rightarrow CaCl_{2(aq)} + CO_{2(g)} + H_2O_{(l)}$$

N.B. When given appropriate information, you must be able to write and balance symbol equations.

Relative Atomic Mass

Atoms are too small for their actual atomic mass to be of much use to us. We therefore use the **relative atomic mass** (**RAM** or A_r). This is a number that compares the mass of one atom to the mass of other atoms.

Each element in the periodic table has two numbers. The larger of the two (at the top of the symbol) is the mass number, which also doubles up as the relative atomic mass.

Example

$^{65}_{30}Zn$
Zinc

Relative atomic mass of zinc is 65.

Relative Formula Mass

The **relative formula mass** (**RFM** or M_r) of a compound is the relative atomic masses of all its elements added together. To calculate RFM we need to know the formula of the compound and the RAM of each of the atoms involved.

Example
Calculate the RFM of water, H_2O.

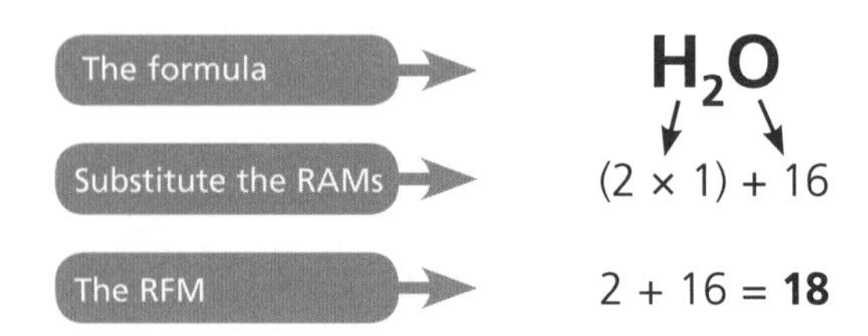

The relative formula mass of water is 18.

Extracting Useful Materials

The lithosphere contains many naturally occurring elements and compounds called **minerals**. Ores are rocks that contain varying amounts of minerals, from which metals can be extracted. Sometimes very large amounts of ores need to be mined in order to recover a small percentage of valuable minerals, e.g. copper. The method of extraction depends on the metal's position in the reactivity series.

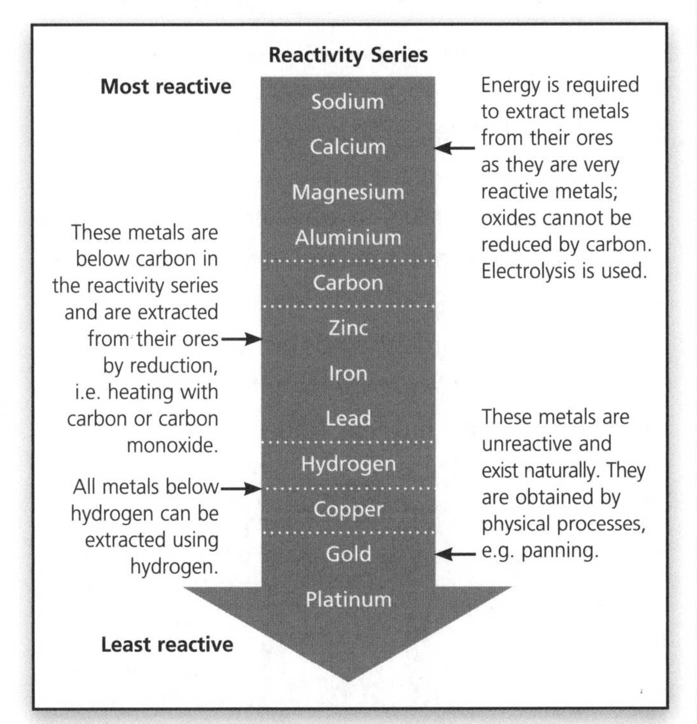

Examples of Extraction by Reduction with Carbon

Zinc can be extracted from zinc oxide by heating it with carbon. Zinc oxide is reduced because it loses oxygen. Carbon is oxidised because it gains oxygen.

Zinc oxide + Carbon → Zinc + Carbon dioxide (Reduction; Oxidation)

Iron and copper can also be extracted using carbon reduction:

Iron oxide + Carbon → Iron + Carbon dioxide (Reduction; Oxidation)

Copper oxide + Carbon → Copper + Carbon dioxide (Reduction; Oxidation)

HT Calculating a Metal's Mass

If you are given its formula, you can calculate the mass of metal that can be extracted from a substance. Follow these steps:

1. Write down the formula.
2. Work out the relative formula mass.
3. Work out the percentage mass of metal in the formula.
4. Work out the mass of metal.

Example

Find the mass of Zn that can be extracted from 100g of ZnO.

1. The formula is ZnO
2. Relative formula mass $= 65 + 16$
   $= 81$
3. Percentage of zinc present
   $= \frac{\text{RAM of Zn}}{\text{RFM of ZnO}} \times 100$
   $= \frac{65}{81} \times 100 = 80\%$
4. In 100g of ZnO, there will be $\frac{80}{100} \times 100$
   = **80g of Zn**

If you were given the equation of a reaction, you could find the ratio of the mass of the reactant to the mass of the product.

$$2ZnO + C \longrightarrow 2Zn + CO_2$$

Relative formula mass:

Work out the RFM of each substance

$(2 \times 81) + 12 = (2 \times 65) + 44$

$162 + 12 = 130 + 44$

$174 = 174$ ✓

Therefore, 162g of ZnO produces 130g of Zn.

So, 1g of ZnO produces: $\frac{130}{162} = 0.8$g of Zn

and 100g of ZnO produces: $0.8 \times 100 =$ **80g of Zn**.

Metals and the Environment

In order to assess the impact on the environment of extracting and using metals, a life cycle assessment of metal products needs to be carried out. You need to be able to understand and evaluate the sort of data in the table below:

Stage of Life Cycle	Process	Environmental Impact
Making the material from natural raw materials	Mining	• Lots of rock wasted • Leaves a scar on the landscape • Air pollution • Noise pollution
	Processing	• Pollutants caused by transportation • Energy usage
	Extracting the metal	• Electrolysis uses more energy than reduction
Making the product from the material	Manufacturing products	• Energy usage in processing and transportation
Use	Transport	• Pollutants caused by transportation
	Running product	• Energy usage
Disposal	Reuse	• No impact
	Recycle	• Uses a lot less energy than the initial manufacturing
	Throw away	• Landfill sites remove wildlife habitats and are unsightly

Extraction by Electrolysis

Electrolysis is the decomposition of an **electrolyte** (a solution that conducts electricity) using an electric current. The process is used in industry to extract reactive metals from their ores. Ionic compounds will only conduct electricity when their ions are free to move. This occurs when the compound is either molten or dissolved in solution. During melting of an ionic compound, the electrostatic forces between the charged ions are broken. The crystal lattice is broken down and the ions are free to move.

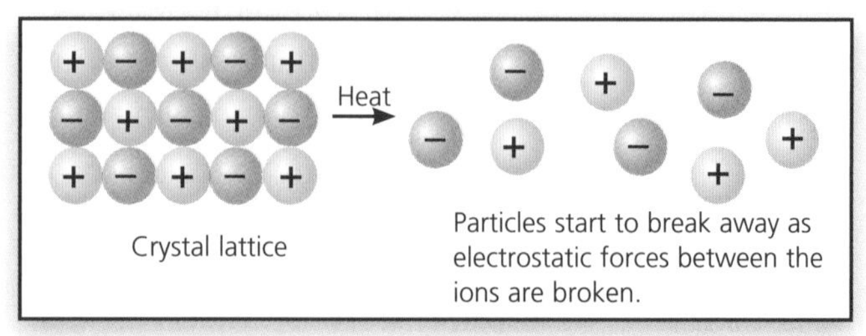

Molten lead bromide is a liquid containing positive lead ions and negative bromide ions that are free to move throughout the liquid. When a direct current is passed through the molten salt, the positively charged lead ions are attracted towards the negative electrode. The negatively charged bromide ions are attracted towards the positive electrode. As a result, lead is formed at the negative electrode and bromine at the positive electrode.

HT

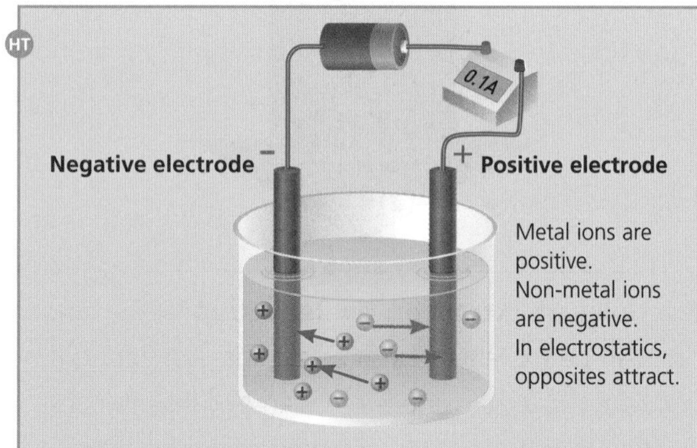

When the ions get to the oppositely charged electrode, they are **discharged** (i.e. they lose their charge).

- The bromide ion loses electrons to the positive electrode to form a bromine atom. The bromine atom then bonds with a second atom to form a bromine molecule.
  $2Br^- \longrightarrow Br_2 + 2e^-$
- The lead ions gain electrons from the negative electrode to form a lead atom.
  $Pb^{2+} + 2e^- \longrightarrow Pb$

This process completes the circuit as the electrons are exchanged at the electrodes.

Extracting Aluminium by Electrolysis

Aluminium must be obtained from its ore by electrolysis because it is too reactive to be extracted by heating with carbon. (Look at its position in the reactivity series.)

The steps in the process are as follows:

1. Aluminium ore (bauxite) is purified to leave aluminium oxide.
2. Aluminium oxide is mixed with cryolite (a compound of aluminium) to lower its melting point.
3. The mixture of aluminium oxide and cryolite is melted so that the ions can move.
4. When a current passes through the molten mixture, positively charged aluminium ions move towards the negative electrode (the cathode), and aluminium is formed. Negatively charged oxide ions move towards the positive electrode (the anode), and oxygen is formed.
5. This causes the positive electrodes to burn away quickly. They have to be replaced frequently.

The electrolysis of bauxite to obtain aluminium is quite an expensive process due to the cost of the large amounts of electrical energy needed to carry it out. The equation for this reaction is:

Aluminium oxide	→	Aluminium	+	Oxygen
$2Al_2O_3(l)$	→	$4Al(l)$	+	$3O_2(g)$

HT The reactions at the electrodes can be written as half equations. This means that we write separate equations for what is happening at each of the electrodes during electrolysis.

At the negative electrode (cathode)...

$$Al^{3+}(l) + 3e^- \xrightarrow{\text{Reduction}} Al(l)$$

At the positive electrode (anode)...

$$2O^{2-}(l) - 4e^- \xrightarrow{\text{Oxidation}} O_2(g)$$

Electrolysis of Aluminium Oxide

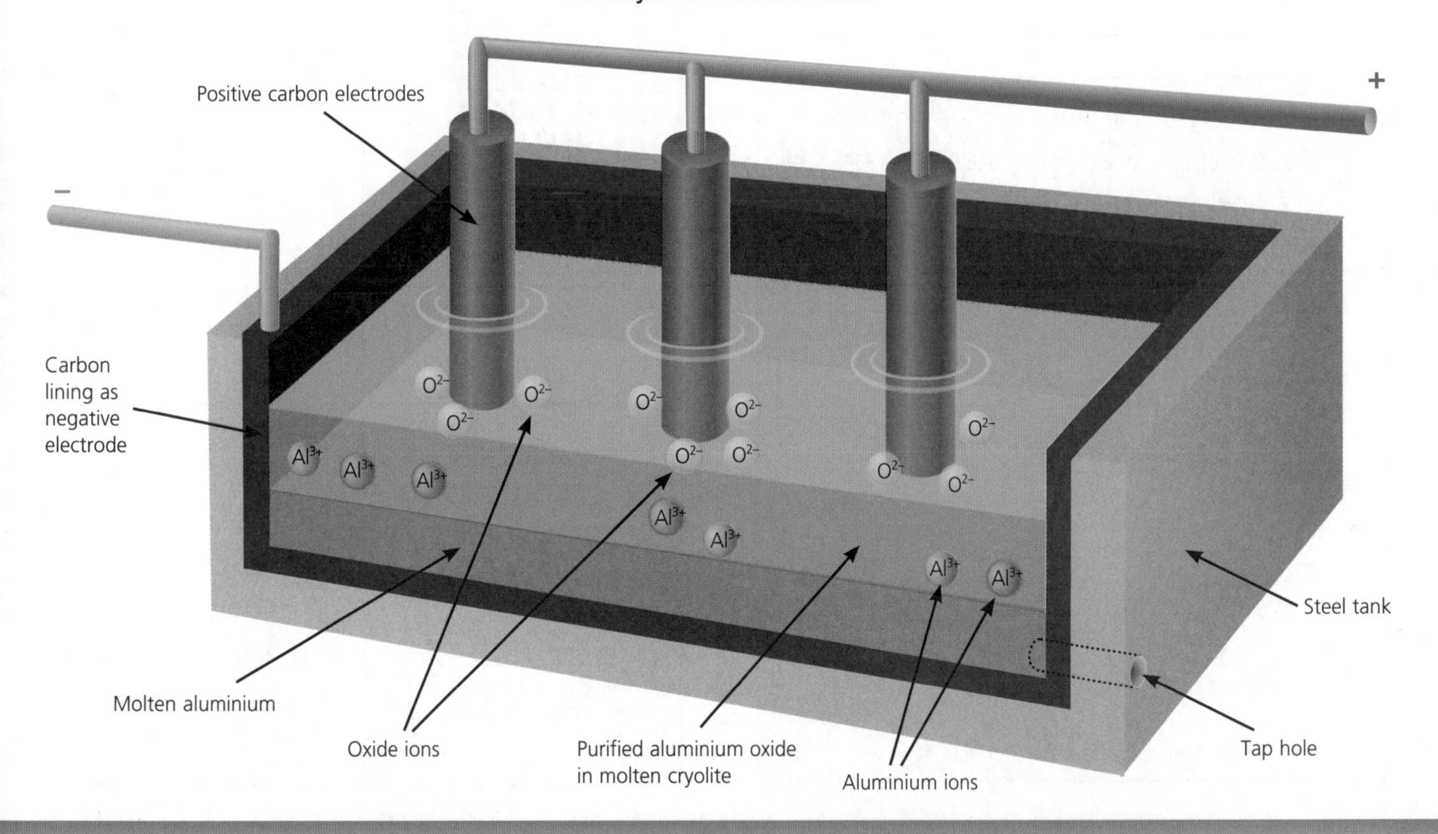

Properties of Metals

Generally, metals are strong and malleable, have high melting points and can conduct electricity. Their properties determine how each particular metal can be used. For example:

Metal	Properties	Uses
Titanium	• Very strong • Lightweight • Resistant to corrosion	• Replacement hip joints • Bicycles • Submarines
Aluminium	• Malleable • Lightweight • Resistant to corrosion	• Drinks cans • Window frames • Saucepans • Aircraft
Iron	• High melting point • Strong	• Saucepans • Cars
Copper	• Conducts electricity • Conducts heat	• Cables, e.g. kettle cable • Electromagnets • Electrical switches • Saucepans

HT In a metal crystalline structure, the positively charged metal ions are held closely together by a 'sea' of electrons that are free to move.

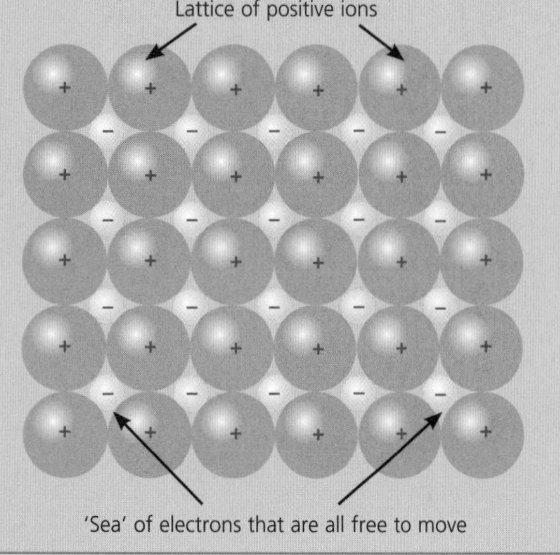

The properties of a metal can be explained by its structure. The force of attraction that keeps the structure together is known as the **metallic bond**. The metallic bond can be used to explain the properties of metals.

Structure	Property and Explanation
The ions are arranged in a lattice form	**Very strong** Metal ions are closely packed in a lattice structure.
	High melting point A lot of energy is needed to break the strong force of attraction between the metal ions and the sea of electrons.
HT Force applied Rows of ions slide over each other. Result: The metal is 'bendy' and can be dented	**Malleable** External forces cause layers of metal ions to move by sliding over other layers.
Moving electrons can carry the electric charge	**Conducts electricity** Electrons are free to move throughout the structure. When an electrical force is applied, the electrons move along the metal in one direction.

Module C6 (Chemical Synthesis)

Chemical synthesis provides the chemicals needed for food processing, health care and many other products. This module looks at:
- chemicals, and why we need them
- acids, alkalis and their reactions
- calculating the mass of products and reactants
- how titrations are used
- measuring the rate of a reaction
- planning, carrying out and controlling a chemical synthesis.

Chemicals

Chemicals are all around us and we depend on them daily. **Chemical synthesis** is the process by which raw materials are made into useful products such as:
- food additives
- fertilisers
- dyestuffs
- pigments
- pharmaceuticals
- cosmetics
- paints.

The chemical industry makes **bulk chemicals**, such as sulfuric acid and ammonia, on a very large scale (millions of tonnes per year). **Fine chemicals**, such as drugs and pesticides, are made on a much smaller scale.

The range of chemicals made in industry and laboratories in the UK is illustrated in the pie chart below.

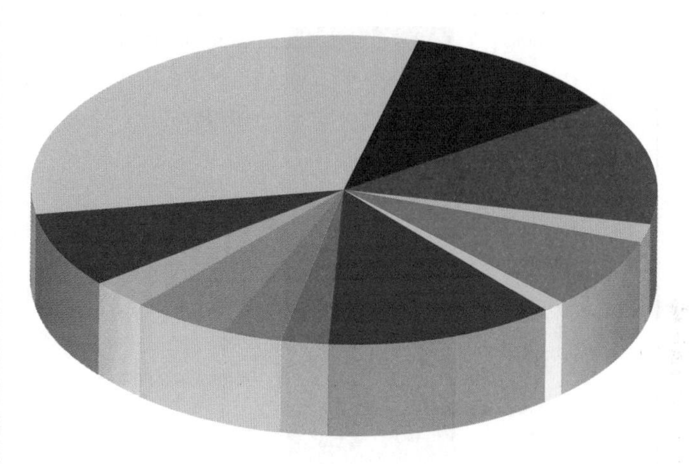

Key:
- Pharmaceuticals 31.5%
- Paint, varnishes and printing inks 8%
- Agrochemicals 3%
- Industrial glass 5%
- Dyes and pigments 3%
- Basic inorganics 2.5%
- Basic organics 12%
- Fertilisers 1%
- Plastic and synthetic rubber 7.5%
- Synthetic fibres 2%
- Other specialities 13%
- Soaps, toiletries and cleaning preparations 11.5%

Many of the raw materials are hazardous, therefore it is important to recognise the standard hazard symbols and understand the necessary precautions that need to be taken (see Module C4).

C6 Chemical Synthesis

The pH Scale

The pH scale is a measure of the acidity or alkalinity of an aqueous solution, across a 14-point scale.

Acids are substances that have a pH less than 7.

Bases are the oxides and hydroxides of metals. Those which are soluble are called **alkalis** and they have a pH greater than 7.

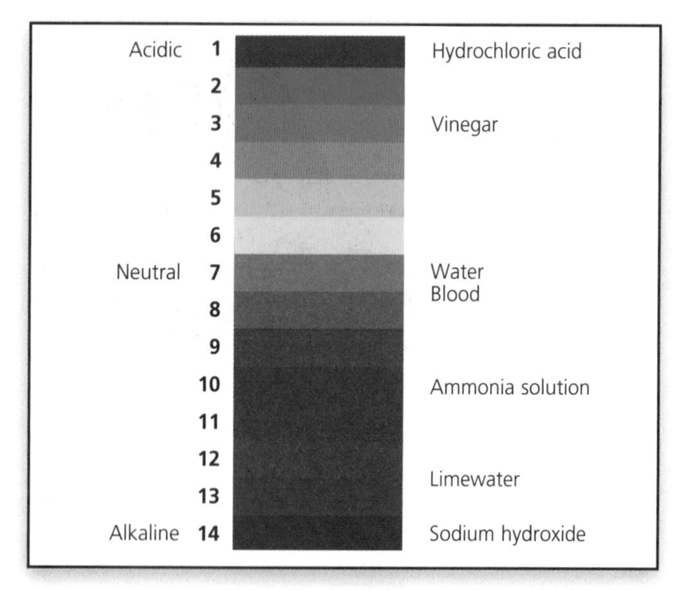

Indicators such as litmus paper, universal indicator and pH meters are used to detect whether a substance is acidic or alkaline. The pH of a substance is measured using a full range indicator, such as universal indicator solution, or a pH meter.

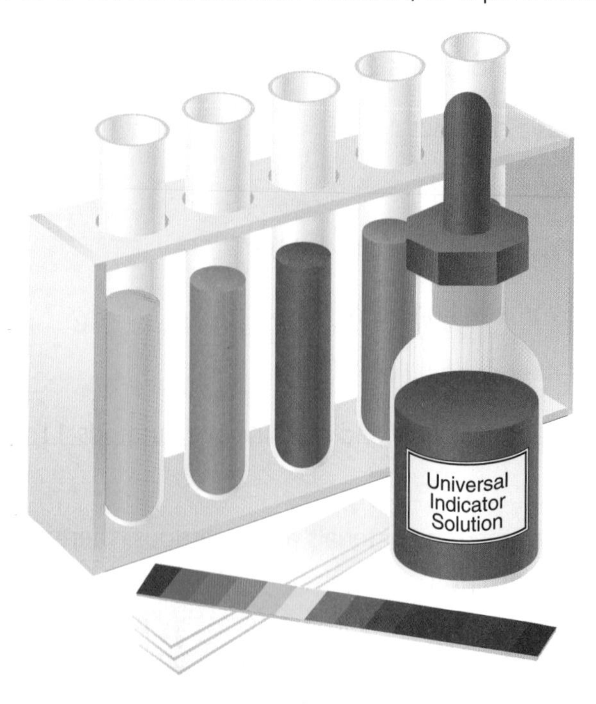

Acidic compounds produce aqueous **hydrogen ions**, $H^+(aq)$, when they dissolve in water.

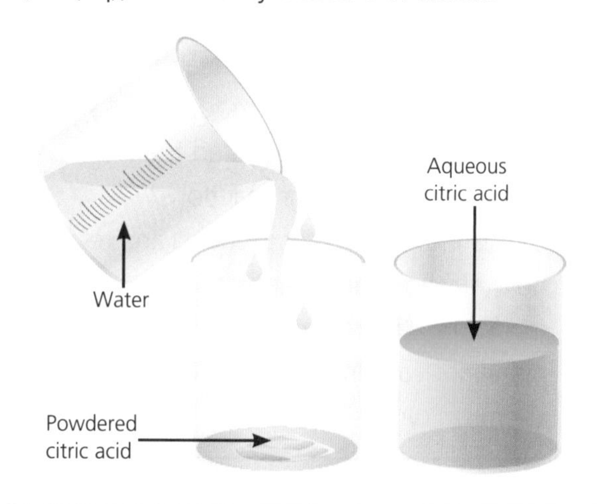

Common Acids	Formulae to Remember	State at Room Temp.
Citric acid	–	Solid
Tartaric acid	–	Solid
Nitric acid	HNO_3	Liquid
Sulfuric acid	H_2SO_4	Liquid
Ethanoic acid	–	Liquid
Hydrogen chloride (hydrochloric acid)	HCl	Gas

Alkalis produce aqueous **hydroxide ions**, $OH^-(aq)$, when they dissolve in water.

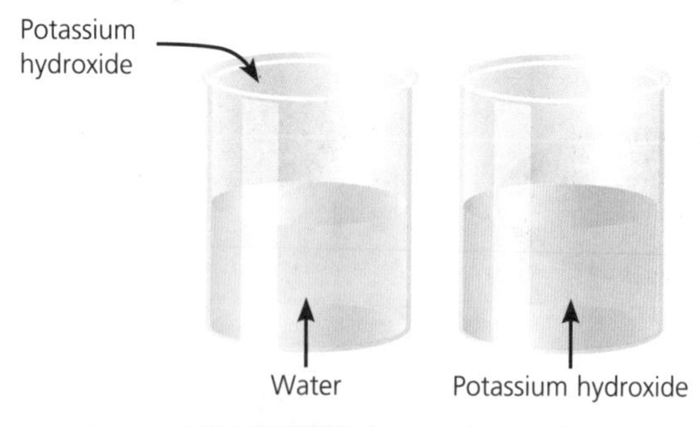

Common Alkalis	Formulae to Remember
Sodium hydroxide	NaOH
Potassium hydroxide	–
Magnesium hydroxide	$Mg(OH)_2$
Calcium hydroxide	–

Neutralisation

When an acid and a base are mixed together in the correct amounts, they 'cancel' each other out. This reaction is called **neutralisation** because the solution that remains has a neutral pH of 7.

Acid + Base → Neutral salt solution + Water

Example

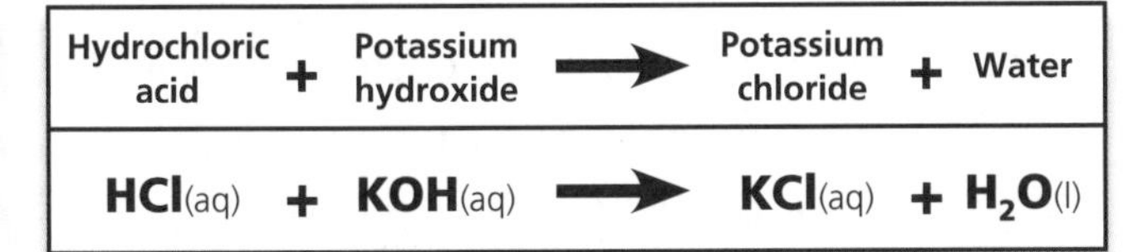

Hydrochloric acid	+	Potassium hydroxide	→	Potassium chloride	+	Water
$HCl_{(aq)}$	+	$KOH_{(aq)}$	→	$KCl_{(aq)}$	+	$H_2O_{(l)}$

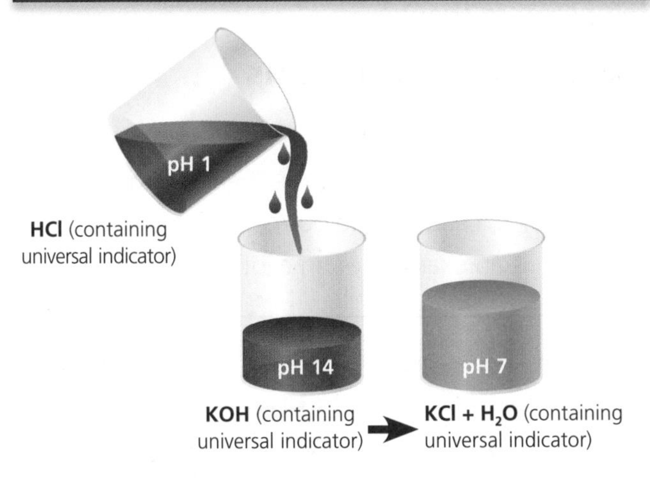

N.B. A balanced equation for a chemical reaction shows the relative numbers of atoms and molecules of reactants and products taking part in the reaction.

During neutralisation, the hydrogen ions from the acid react with the hydroxide ions from the alkali to make water. The simplest way of writing a neutralisation equation is:

$$H^+_{(aq)} + OH^-_{(aq)} \rightarrow H_2O_{(l)}$$

You need to remember this

The salt produced during neutralisation depends on the metal in the base and the acid used. Hydrochloric acid produces chloride salts; sulfuric acid produces sulfate salts; and nitric acid produces nitrate salts.

HT You must be able to write balanced equations to describe the characteristic reactions of acids.

Hydrochloric Acid

Hydrochloric acid	+	Zinc	→	Zinc chloride	+	Hydrogen
$2HCl_{(aq)}$	+	$Zn_{(s)}$	→	$ZnCl_{2(aq)}$	+	$H_{2(g)}$

Hydrochloric acid	+	Potassium hydroxide	→	Potassium chloride	+	Water
$HCl_{(aq)}$	+	$KOH_{(aq)}$	→	$KCl_{(aq)}$	+	$H_2O_{(l)}$

Hydrochloric acid	+	Copper oxide	→	Copper chloride	+	Water
$2HCl_{(aq)}$	+	$CuO_{(s)}$	→	$CuCl_{2(aq)}$	+	$H_2O_{(l)}$

Hydrochloric acid	+	Calcium carbonate	→	Calcium chloride	+	Water	+	Carbon dioxide
$2HCl_{(aq)}$	+	$CaCO_{3(s)}$	→	$CaCl_{2(aq)}$	+	$H_2O_{(l)}$	+	$CO_{2(g)}$

Sulfuric Acid

Sulfuric acid	+	Zinc	→	Zinc sulfate	+	Hydrogen
$H_2SO_{4(aq)}$	+	$Zn_{(s)}$	→	$ZnSO_{4(aq)}$	+	$H_{2(g)}$

Sulfuric acid	+	Potassium hydroxide	→	Potassium sulfate	+	Water
$H_2SO_{4(aq)}$	+	$2KOH_{(aq)}$	→	$K_2SO_{4(aq)}$	+	$2H_2O_{(l)}$

Sulfuric acid	+	Copper oxide	→	Copper sulfate	+	Water
$H_2SO_{4(aq)}$	+	$CuO_{(s)}$	→	$CuSO_{4(aq)}$	+	$H_2O_{(l)}$

Sulfuric acid	+	Calcium carbonate	→	Calcium sulfate	+	Water	+	Carbon dioxide
$H_2SO_{4(aq)}$	+	$CaCO_{3(s)}$	→	$CaSO_{4(aq)}$	+	$H_2O_{(l)}$	+	$CO_{2(g)}$

Nitric Acid

Nitric acid	+	Zinc	→	Zinc nitrate	+	Hydrogen
$2HNO_{3(aq)}$	+	$Zn_{(s)}$	→	$Zn(NO_3)_{2(aq)}$	+	$H_{2(g)}$

Nitric acid	+	Potassium hydroxide	→	Potassium nitrate	+	Water
$HNO_{3(aq)}$	+	$KOH_{(aq)}$	→	$KNO_{3(aq)}$	+	$H_2O_{(l)}$

Nitric acid	+	Copper oxide	→	Copper nitrate	+	Water
$2HNO_{3(aq)}$	+	$CuO_{(s)}$	→	$Cu(NO_3)_{2(aq)}$	+	$H_2O_{(l)}$

Nitric acid	+	Calcium carbonate	→	Calcium nitrate	+	Water	+	Carbon dioxide
$2HNO_{3(aq)}$	+	$CaCO_{3(s)}$	→	$Ca(NO_3)_{2(aq)}$	+	$H_2O_{(l)}$	+	$CO_{2(g)}$

C6 Chemical Synthesis

Writing Formulae

You need to remember the formulae of the salts listed in this table.

Group of Metal	Salt	Formula to Remember
1	Sodium chloride	$NaCl$
1	Potassium chloride	KCl
1	Sodium carbonate	Na_2CO_3
1	Sodium nitrate	$NaNO_3$
1	Sodium sulfate	Na_2SO_4
2	Magnesium sulfate	$MgSO_4$
2	Magnesium carbonate	$MgCO_3$
2	Magnesium oxide	MgO
2	Magnesium chloride	$MgCl_2$
2	Calcium carbonate	$CaCO_3$
2	Calcium chloride	$CaCl_2$

HT You should already know how to write formulae for **ionic compounds**. Given the formulae of the salts listed in the table above, you need to be able to work out the charge on each ion in a compound. See Module C4.

It is important to know the formulae of the common gases that occur as covalently bonded (diatomic) molecules.

Gas	Formula to Remember
Chlorine	Cl_2
Hydrogen	H_2
Nitrogen	N_2
Oxygen	O_2

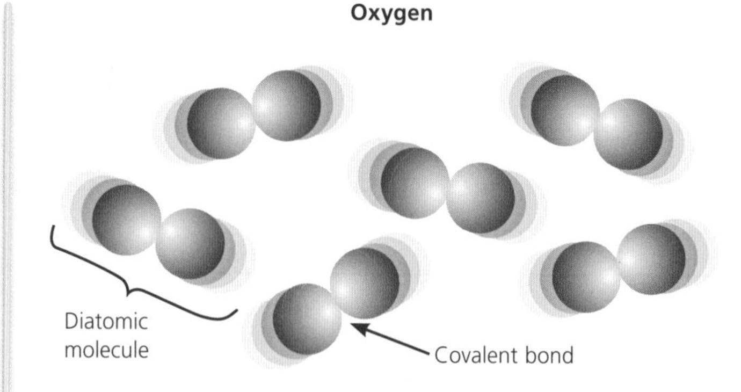

Energy Changes

When chemical reactions occur, energy is transferred to or from the surroundings. Therefore, many chemical reactions are accompanied by a **temperature change**.

Exothermic Reactions

Exothermic reactions are accompanied by a **temperature rise**. They transfer heat energy to the surroundings, i.e. they give out heat. The combustion of carbon is an example of an exothermic reaction:

Carbon	+	Oxygen	⟶	Carbon dioxide
C	+	O_2	⟶	CO_2

It is not only reactions between fuels and oxygen that are exothermic. Neutralising alkalis with acids and many oxidation reactions also give out heat.

The energy change in an exothermic reaction can be shown using an energy-level diagram. Energy is lost to the surroundings during the reaction, so the products have less energy than the reactants.

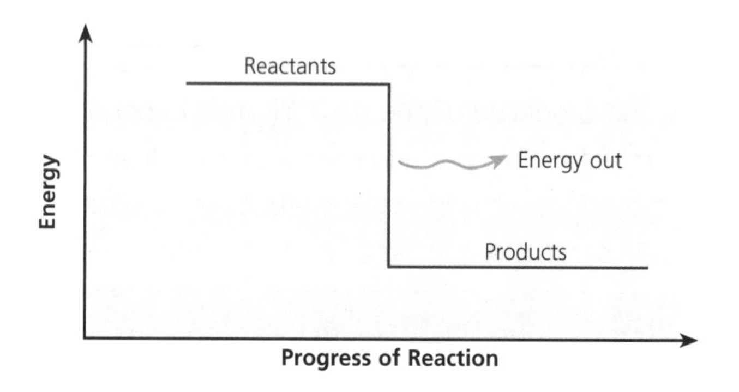

Endothermic Reactions

Endothermic reactions are accompanied by a **fall in temperature**. Heat is transferred from the surroundings, i.e. they take in heat. The reaction between citric acid and sodium hydrogencarbonate is an example of an endothermic chemical reaction. Dissolving ammonium nitrate crystals in water is an example of a physical change that is endothermic.

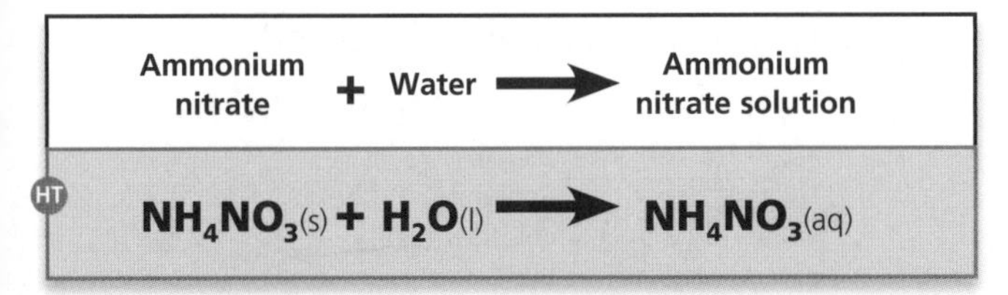

Thermal decomposition is also an example of an endothermic reaction.

The energy change in an endothermic reaction can be shown using an energy-level diagram. Energy is taken in during the reaction, so the products have more energy than the reactants.

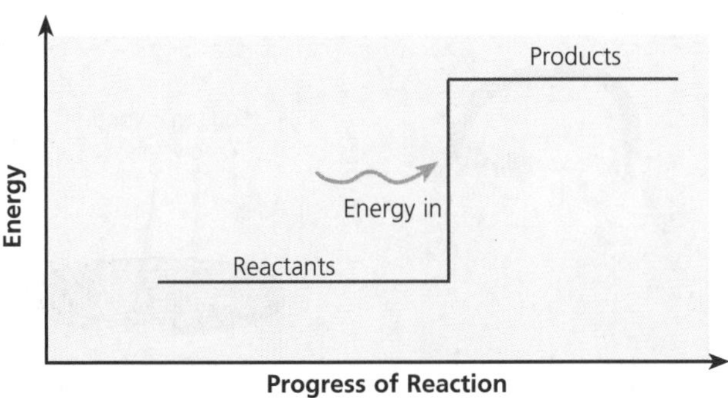

Since energy changes play a major role in chemical reactions, it is very important that these changes are managed during chemical synthesis. For example, there could be a nasty accident if too much energy was given out during a reaction.

Chemical Synthesis

Whenever chemical synthesis takes place, the starting materials (reactants) react to produce new substances (products). The greater the amount of reactants used, the greater the amount of product formed. The **percentage yield** can be calculated by comparing the actual amount of product made (actual yield) with the amount of product you would expect to get if the reaction goes to completion (theoretical yield).

$$\text{Percentage yield} = \frac{\text{Actual yield}}{\text{Theoretical yield}} \times 100$$

There are a number of different stages to any chemical synthesis of an inorganic compound. Look at the flow chart below.

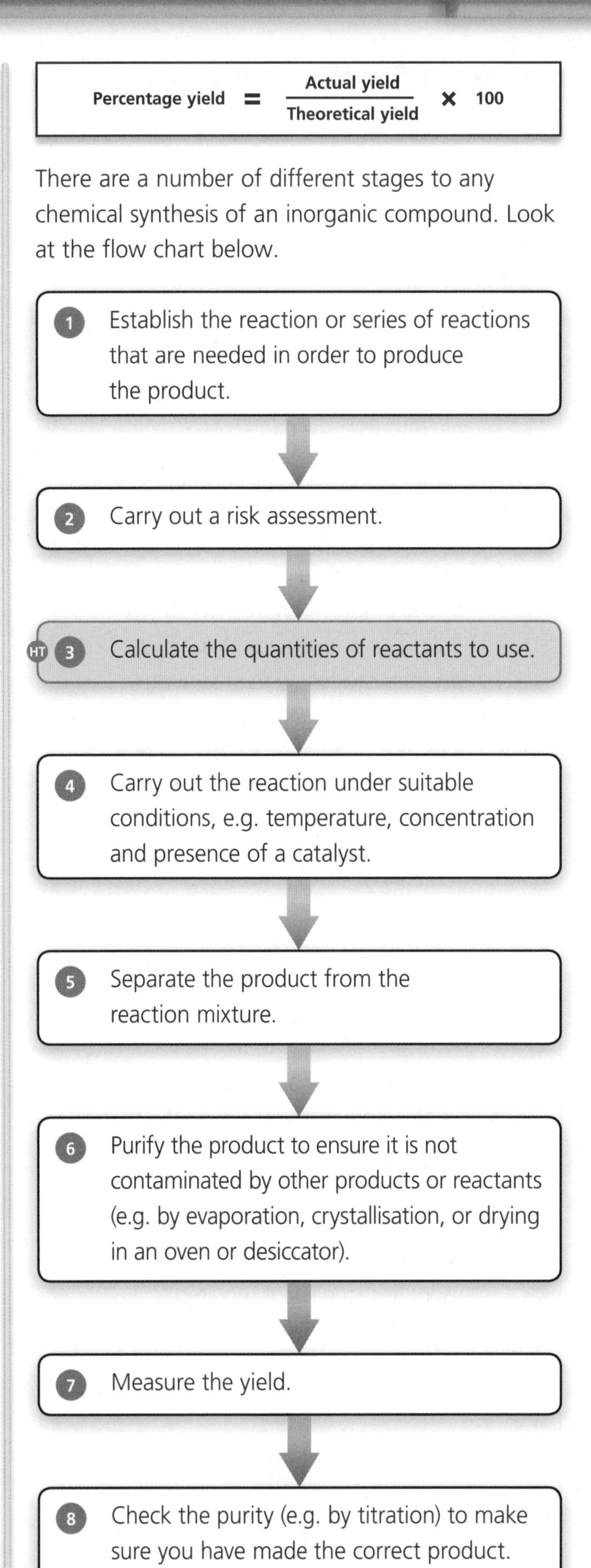

Chemical Synthesis (Cont.)

Example

The following steps show how to make **sodium chloride**.

1. $\mathbf{NaOH}_{(aq)} + \mathbf{HCl}_{(aq)} \longrightarrow \mathbf{NaCl}_{(aq)} + \mathbf{H_2O}_{(l)}$
2. Wear safety glasses and use dilute hydrochloric acid.

HT 3. To calculate the quantities of reactants you need to use, see page 57.

4.
- Measure out $25cm^3$ of hydrochloric acid in a measuring cylinder and pour it into a beaker.
- Add a few drops of indicator, such as phenolphthalein, which will stay colourless unless alkali is present. Alternatively, use a pH meter to follow the reaction.
- Slowly add sodium hydroxide until the indicator changes or the pH meter reads 7. Pink is the alkali colour for phenolphthalein.

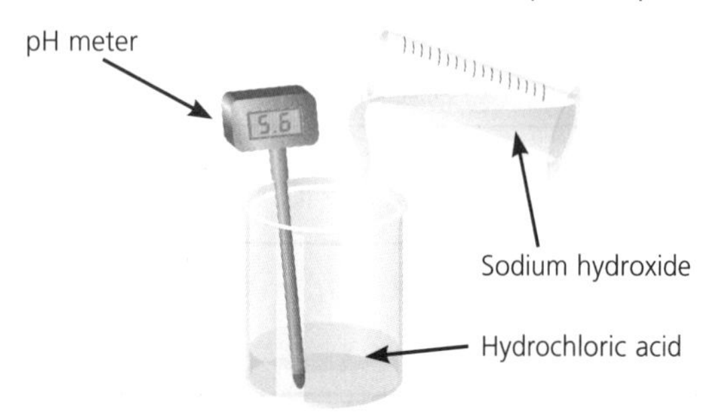

- Mix a spatula of activated charcoal (charcoal treated with oxygen) into the solution. The indicator molecules will be absorbed onto the surface of the charcoal. Filter to remove both the activated charcoal and the indicator. Note, if a pH meter was used, this step is not required.

Additional Step to Remove Indicator

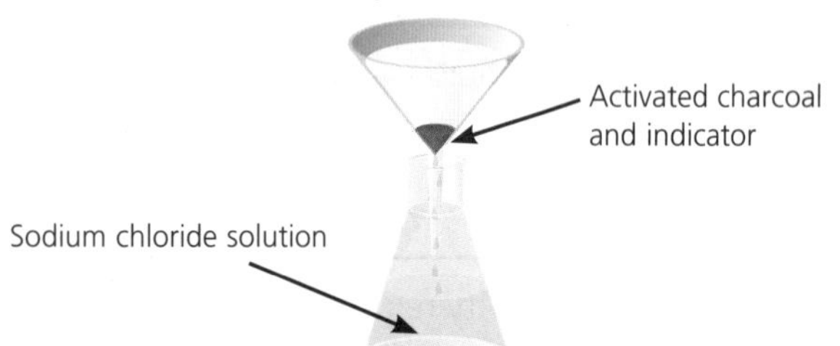

5. Remove the water by gently heating the solution. It will evaporate slowly to form crystals that will cling to the end of a cold glass rod. Leave them to cool and crystallise. Filter to separate the crystals from any solution left behind.

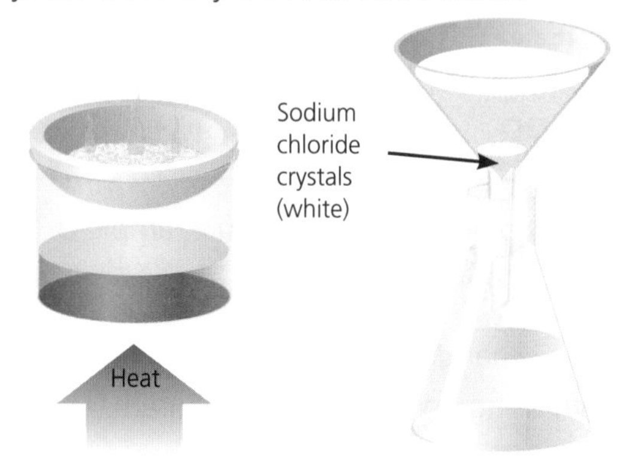

6. Wash the crystals to remove any excess NaOH and dry in a desiccator or oven.

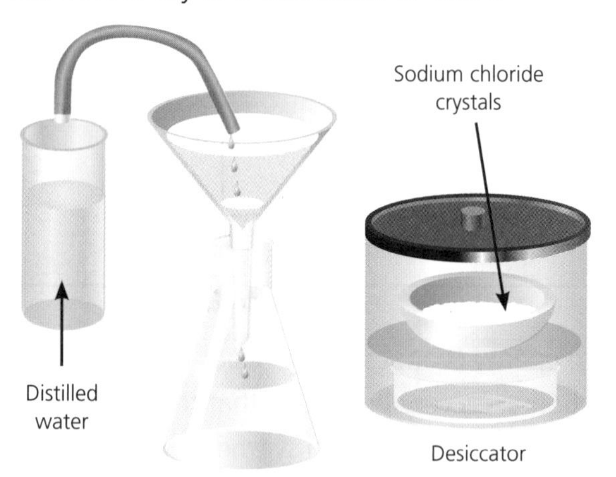

7. Weigh the mass and calculate the percentage yield.

$$\text{Percentage yield} = \frac{\text{Actual yield}}{\text{Theoretical yield}} \times 100$$

Actual yield = 8g

Theoretical yield = 10g

$$\text{Percentage yield} = \frac{8}{10} \times 100 = \mathbf{80\%}$$

8. Check the purity by titration (see page 58).

Quantity of Reactants and Products

In order to work out how much of each reactant is required to make a known amount of product, you must understand:

- that a balanced equation shows the relative number of atoms or molecules of reactants and products taking part in the reaction
- that the relative atomic mass of an element shows the mass of its atoms relative to the mass of other atoms
- that relative atomic masses of elements can be found from the periodic table
- how to calculate the relative atomic mass (see page 46)
- how to substitute the relative formula masses and data into a given mathematical formula to calculate the reacting masses and/or products from a chemical reaction

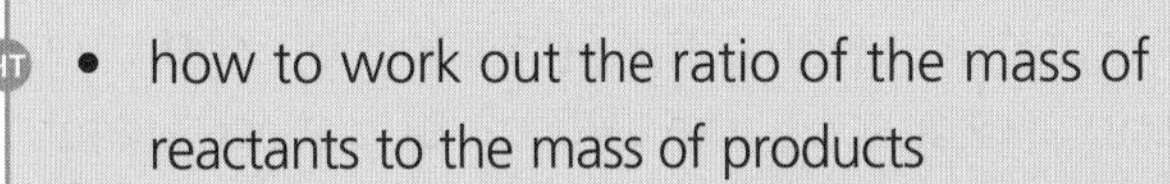

HT

- how to work out the ratio of the mass of reactants to the mass of products
- how to apply the ratio to the question.

HT Calculating a Product's Mass

Example

Calculate how much calcium oxide can be produced from 50kg of calcium carbonate. (Relative atomic masses: Ca = 40, C = 12, O = 16).

Write down the equation.

$CaCO_3 \rightarrow CaO + CO_2$

Work out the RFM of each substance.

$40 + 12 + (3 \times 16) \rightarrow (40 + 16) + [12 + (2 \times 16)]$

Check that the total mass of the reactants equals the total mass of the products. If they are not the same, check your work.

$100 \rightarrow 56 + 44$ ✓

Since the question only mentions calcium oxide and calcium carbonate, you can now ignore the carbon dioxide. You just need the ratio of mass of reactant to mass of product.

100 : 56

We need to know how much CaO is produced from 50kg of $CaCO_3$.

100kg of $CaCO_3$ produces 56kg of CaO.

So, 1kg of $CaCO_3$ produces $\frac{56}{100}$ kg of CaO,

and 50kg of $CaCO_3$ produces $\frac{56}{100} \times 50$

= **28kg** of CaO

Calculating a Reactant's Mass

Example

Calculate how much aluminium oxide is needed to produce 540 tonnes of aluminium. (Relative atomic masses: Al = 27, O = 16).

Write down the equation.

$2Al_2O_3 \rightarrow 4Al + 3O_2$

Work out the RFM of each substance.

$2[(2 \times 27) + (3 \times 16)] \rightarrow (4 \times 27) + [3 \times (2 \times 16)]$

Check that the total mass of the reactants equals the total mass of the products.

$204 \rightarrow 108 + 96$ ✓

Since the question only mentions aluminium oxide and aluminium, you can now ignore the oxygen. You just need the ratio of mass of reactant to mass of product.

204 : 108

We need to know how much Al_2O_3 is needed to produce 540 tonnes of Al.

204 tonnes of Al_2O_3 produces 108 tonnes of Al.

So, $\frac{204}{108}$ tonnes is needed to produce 1 tonne of Al, and $\frac{204}{108} \times 540$ tonnes is needed to produce 540 tonnes of Al

= **1020 tonnes** of Al_2O_3

Titration

Titration can be used to calculate the concentration of an acid, such as citric acid, by finding out how much alkali is needed to neutralise it.

1. Fill a burette with the alkali sodium hydroxide (the concentration of the alkali must be known) and take an initial reading of the volume.
2. Accurately weigh out a 4g sample of solid citric acid and dissolve it in $100cm^3$ of distilled water.
3. Use a pipette to measure $25cm^3$ of the aqueous citric acid and put it into a conical flask.

   Add a few drops of the indicator, phenolphthalein, to the conical flask. The indicator will stay colourless.

   Place the flask on a white tile under the burette.
4. Add the alkali from the burette to the acid in the flask drop by drop.
   - Swirl the flask to ensure it mixes well. Near the end of the reaction, the indicator will start to turn pink.
   - Keep swirling and adding the alkali until the indicator is completely pink, showing that the citric acid has been neutralised.
   - Record the final burette reading. Work out the volume of alkali added from:
     Volume = final reading – initial reading

Repeat the whole procedure until you get two results that are within +/– $0.05cm^3$.

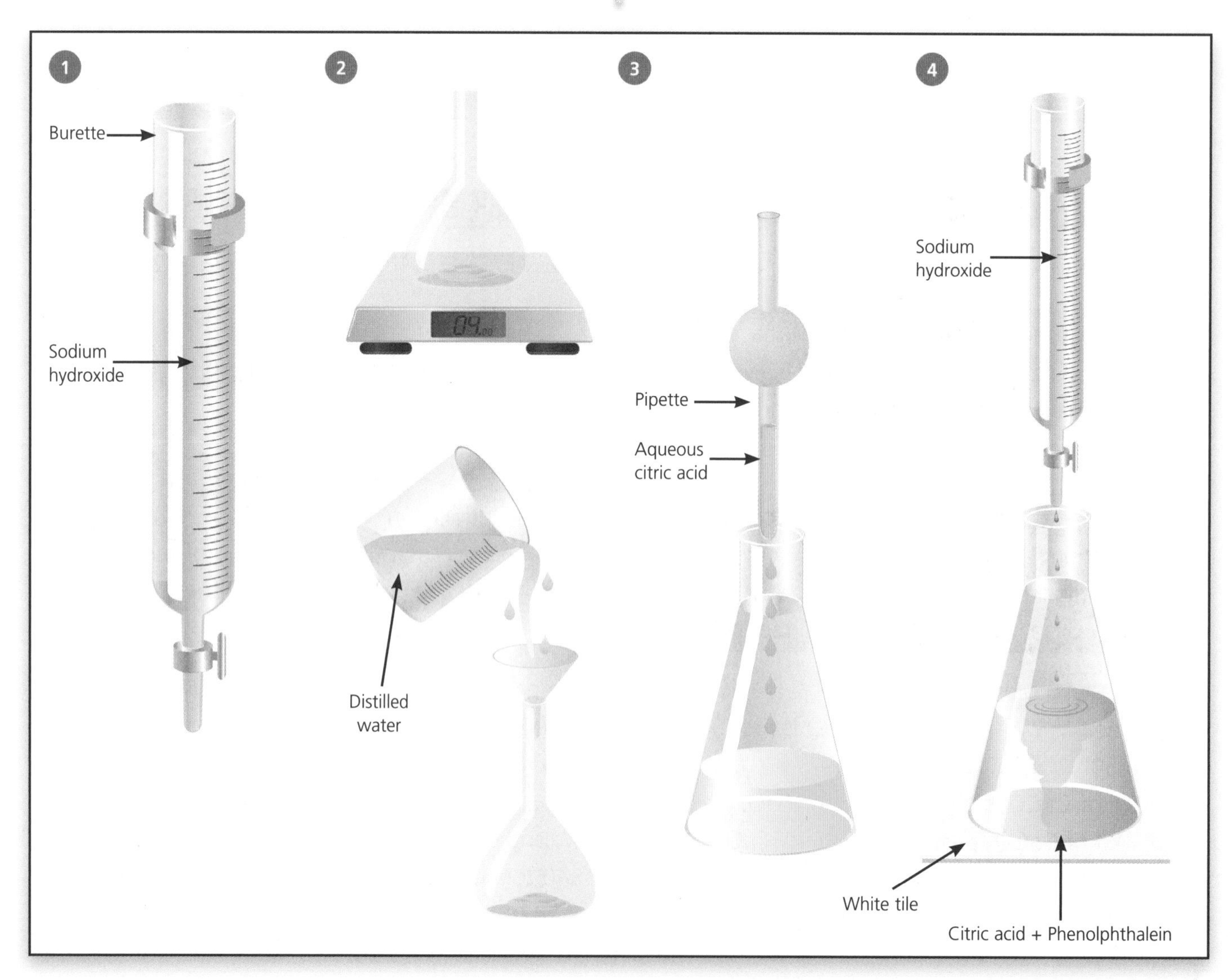

Interpreting Results

Example

Calculate the purity of citric acid used when:

- concentration of sodium hydroxide (NaOH) = $40.0 g/dm^3$
- volume of sodium hydroxide = $8.0 cm^3$
- mass of citric acid = 4.0g
- volume of citric acid solution = $25.0 cm^3$

First, using the formula below, calculate the concentration of citric acid by substituting the values.

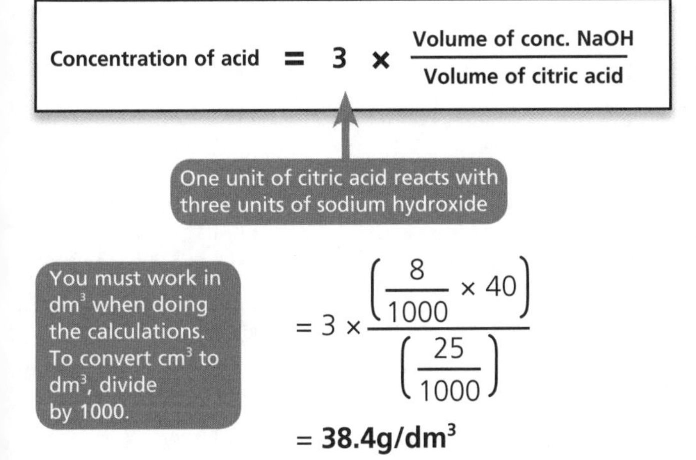

Then, work out the actual mass of citric acid in the sample:

$$\textbf{Mass} = \textbf{Concentration} \times \textbf{Volume}$$

$$= 38.4 g/dm^3 \times \left(\frac{25 cm^3}{1000}\right)$$

$$= \mathbf{0.96g}$$

If the mass of citric acid dissolved in $25 cm^3$ is 0.96g, the mass of citric acid dissolved in $100 cm^3$ of water will be $4 \times 0.96 = 3.84g$

$$\textbf{\% of Purity} = \frac{\textbf{Calculated mass}}{\textbf{Mass weighed out at start}} \times \textbf{100}$$

$$= \frac{3.84}{4.0} \times 100$$

$$= \mathbf{96\%}$$

N.B. You must be able to substitute the results into a formula to interpret them.

Rates of Reactions

The rate of a **chemical reaction** is the amount of reaction that takes place in a given unit of time. Chemical reactions only occur when the reacting particles collide with each other with sufficient energy to react.

Chemical reactions can proceed at different speeds, e.g. rusting is a slow reaction, whereas burning is a fast reaction.

Measuring the Rate of Reaction

The rate of a chemical reaction can be found in different ways:

1. **Weighing the reaction mixture.**
   If one of the products is a gas, you could weigh the reaction mixture at timed intervals. The mass of the mixture will decrease.

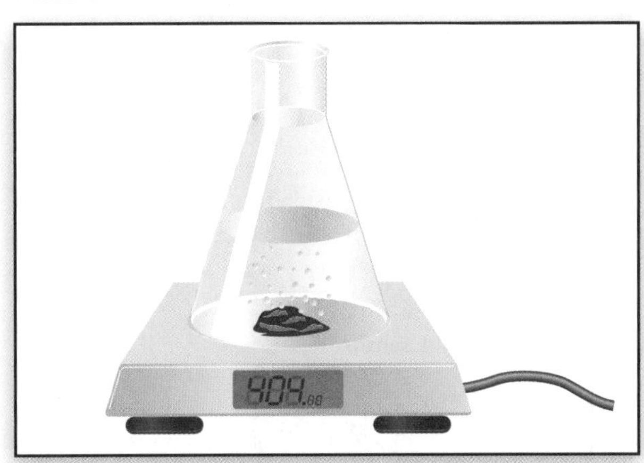

2. **Measuring the volume of gas produced.**
   You could use a gas syringe to measure the total volume of gas produced at timed intervals.

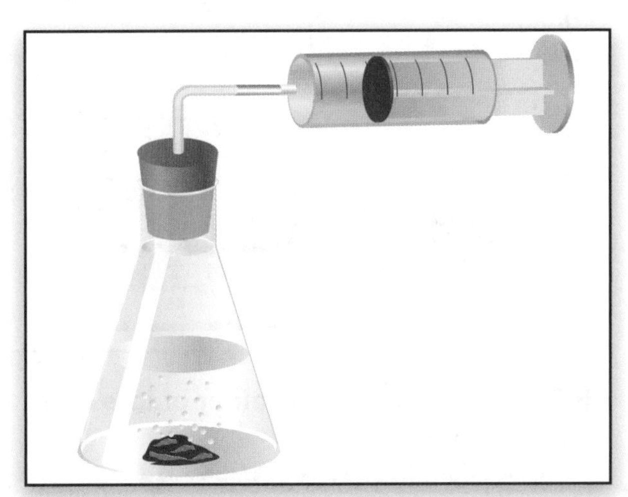

3 Observing the formation of a precipitate.
This can be done by either watching a cross (on a tile underneath the conical flask to see when the cross disappears) in order to measure the formation of a precipitate, or by monitoring a colour change using a light sensor. The light sensor will lead to more reliable and accurate results as there is a definite end point. There are also more data points collected, especially if it is interfaced with a computer.

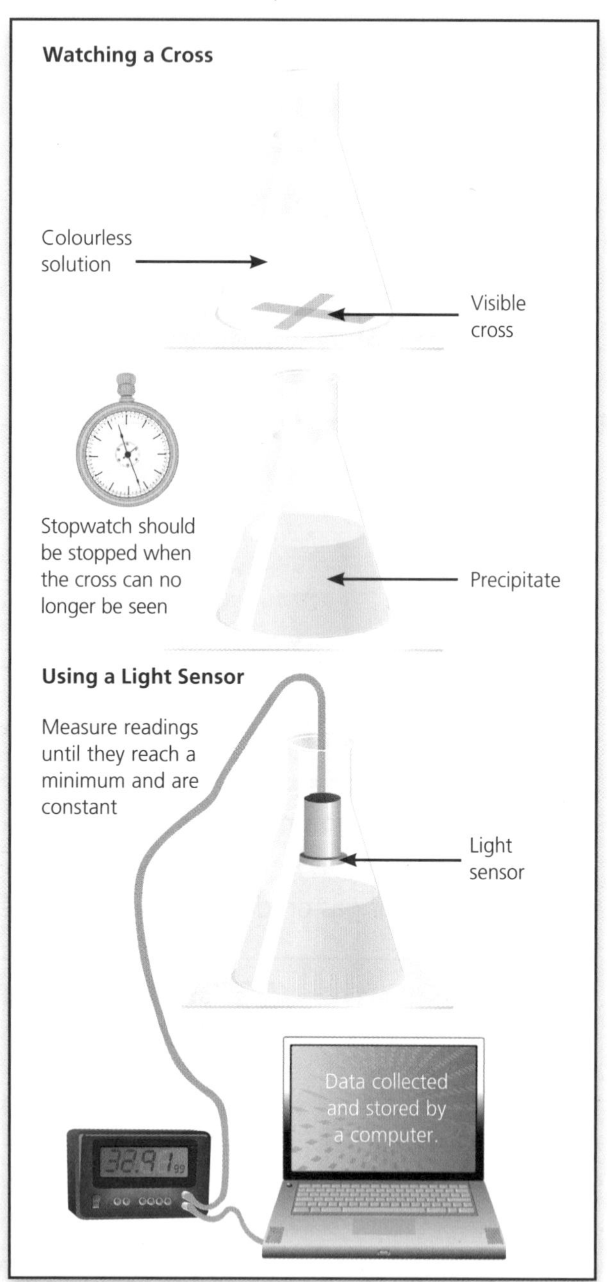

4 Observing the loss of colour or loss of a precipitate.
This is essentially the opposite of 3.

Analysing the Rate of Reaction

Graphs can be plotted to show the progress of a chemical reaction – there are three things to remember:

1. The steeper the line, the faster the reaction.
2. When one of the reactants is used up, the reaction stops (line becomes flat).
3. The same amount of product is formed from the same amount of reactants, irrespective of rate.

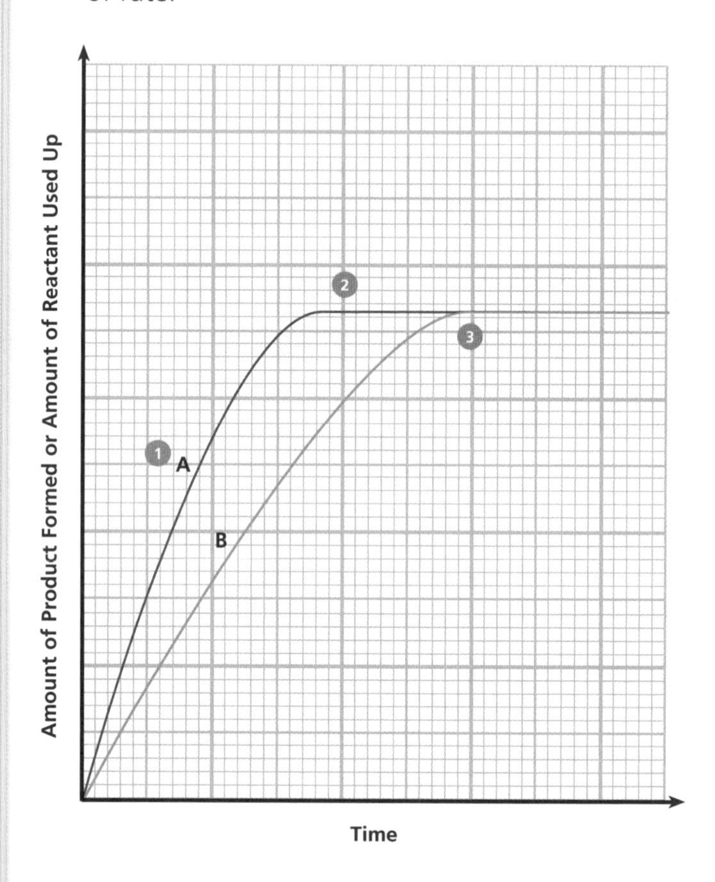

Reaction A is faster than reaction B. This could be because:

- the surface area of the solid reactants in reaction A is greater than in reaction B (i.e. smaller particles are used in A)
- the temperature of reaction A is greater than reaction B
- the concentration of the solution in reaction A is greater than in reaction B
- a catalyst is used in reaction A but not in reaction B.

Changing the Rate of Reaction

There are four important factors that affect the rate of reaction – temperature, concentration of dissolved reactants, surface area and the use of a catalyst.

1 Temperature of the Reactants

In a cold reaction mixture, the particles move quite slowly. They will collide less often, with less energy, so fewer collisions will be successful.

In a hot reaction mixture, the particles move more quickly. They will collide more often, with greater energy, so many more collisions will be successful.

Low Temperature | High Temperature

2 Concentration of the Dissolved Reactants

In a low concentration reaction, the particles are spread out. The particles will collide with each other less often, resulting in fewer successful collisions.

In a high concentration reaction, the particles are crowded close together. The particles will collide with each other more often, resulting in many more successful collisions.

Low Concentration | High Concentration

3 Surface Area of Solid Reactants

Large particles (e.g. solid lumps) have a small surface area in relation to their volume, meaning fewer particles are exposed and available for collisions. This means that particles will collide with each other less often, resulting in fewer successful collisions and therefore a slow rate of reaction.

Small particles (e.g. powdered solids) have a large surface area in relation to their volume, so more particles are exposed and available for collisions. This means that particles will collide with each other more often, resulting in many more successful collisions and therefore a faster rate of reaction.

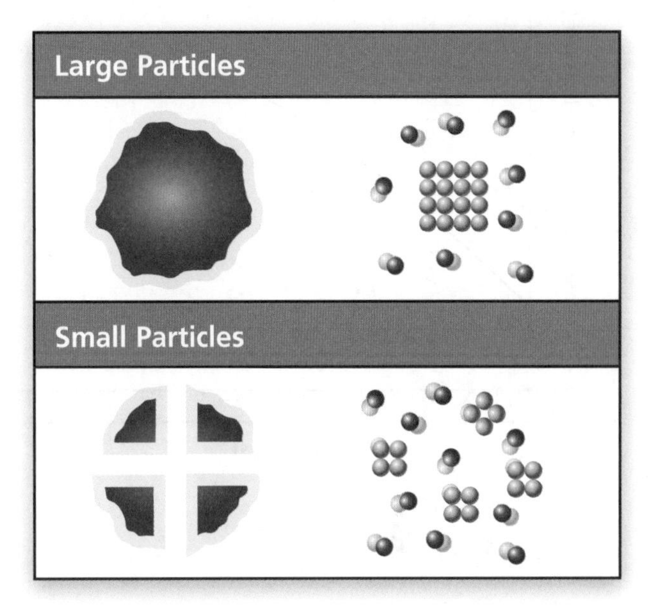

4 Using a Catalyst

A **catalyst** is a substance that increases the rate of a chemical reaction, without being changed during the process. Consider the decomposition of hydrogen peroxide:

Hydrogen peroxide	→	Water	+	Oxygen
$2H_2O_2(aq)$	→	$2H_2O(l)$	+	$O_2(g)$

We can measure the rate of this reaction by measuring the amount of oxygen given off at one-minute intervals. This reaction happens very slowly unless we add a catalyst of manganese(IV) oxide. With a catalyst, plenty of fizzing can be seen as the oxygen is given off.

Changing the Rate of Reaction (Cont.)

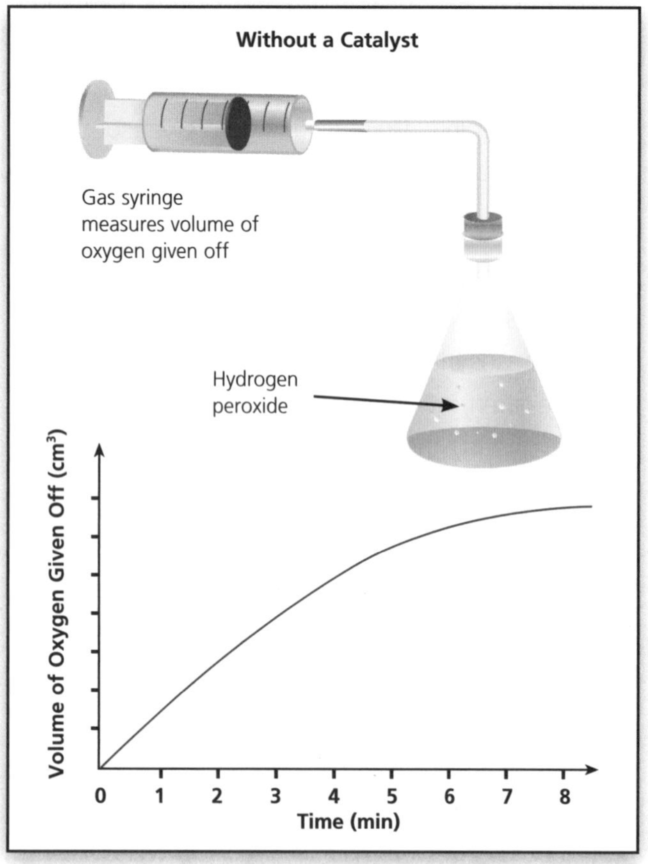

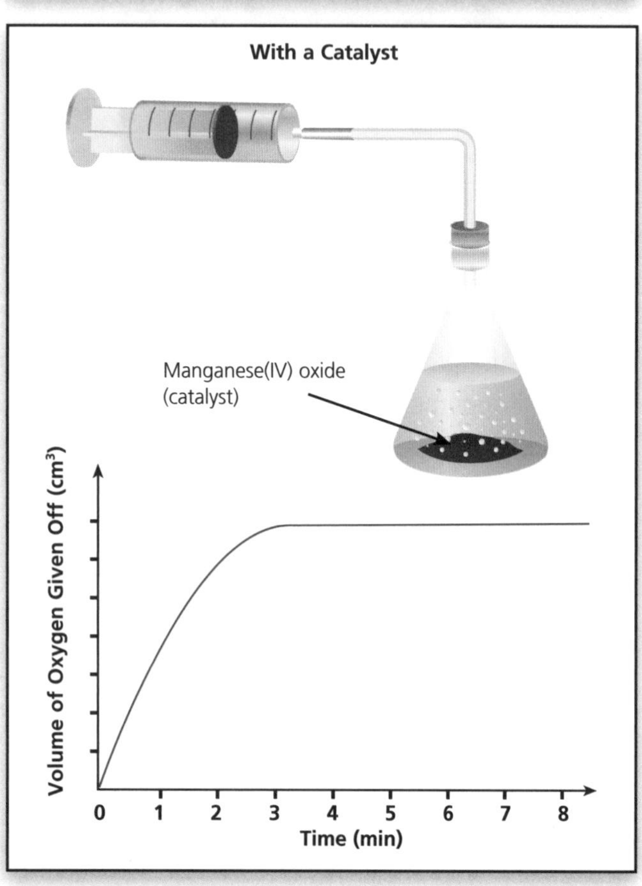

The same amount of gas is given off, but it takes a far shorter time when a catalyst is present.

Catalysts work by lowering the amount of energy needed for a successful collision. They are specific to a particular reaction and are not used up during the reaction. Consequently, only small amounts are needed.

Collision Theory

Chemical reactions only occur when particles collide with each other with sufficient energy. Increasing temperature causes an increase in the kinetic energy of the particles, i.e. they move a lot faster. This results in more energetic collisions happening more frequently. The minimum energy required for a reaction will, therefore, be achieved more often, resulting in a greater rate of reaction.

An increase in concentration or surface area results in more frequent collisions and, therefore, more collisions that are sufficiently energetic for a reaction to occur.

Controlling a Chemical Reaction

When carrying out a chemical synthesis on an industrial scale, there are also economic, safety and environmental factors to consider:

- The rate of manufacture must be high enough to produce a sufficient daily yield of product.
- Percentage yield must be high enough to produce a sufficient daily yield of product.
- A low percentage yield is acceptable, providing the reaction can be repeated many times with recycled starting materials.
- Optimum conditions should be used that give the lowest cost rather than the fastest reaction or highest percentage yield.
- Care must be taken when using any reactants or products that could harm the environment if there was a leak.
- Care must be taken to avoid putting any harmful by-products into the environment.
- A complete risk assessment must be carried out, and the necessary precautions taken.

Exam Practice Questions

1. These diagrams show the atoms in carbon dioxide gas, nitrogen gas and hydrogen gas.

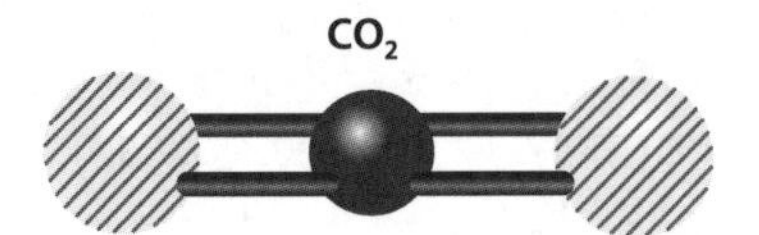

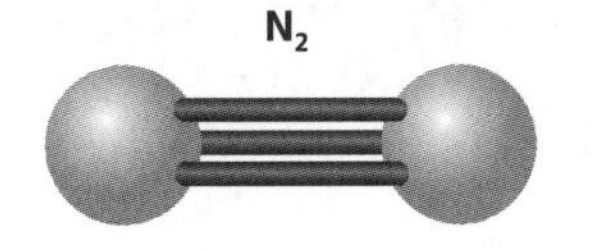

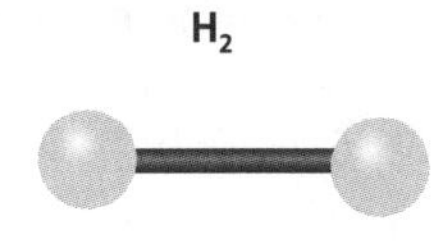

(a) Write down the formulae of the following compounds. **[2]**

(i)

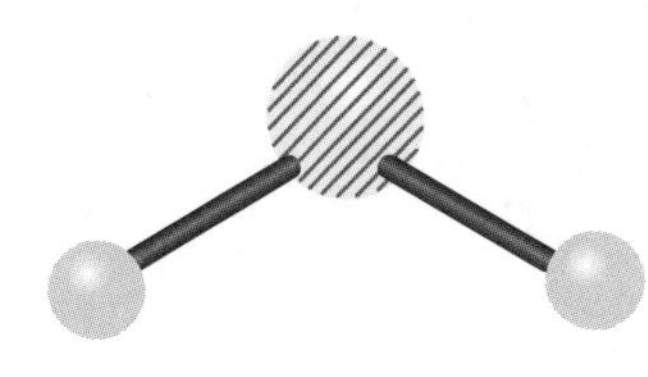

(ii)

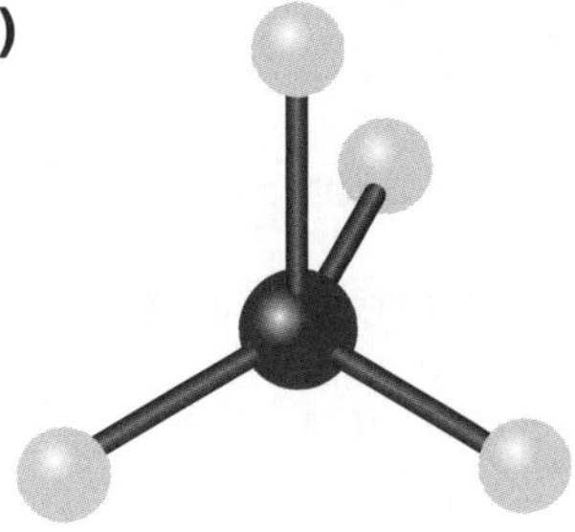

(b) The relative atomic mass of carbon is 12 and the relative atomic mass of oxygen is 16.
What is the relative formula mass of carbon dioxide? **[2]**

2. The lithosphere contains many useful naturally occurring minerals. The process used to extract a metal from its ore depends upon the reactivity of the ore.

(a) Which process best describes the process used to extract iron from its ore?
Put a (ring) around the correct answer. **[1]**

Panning **Reduction** **Electrolysis** **Oxidation**

(b) Before a metal product is manufactured, an analysis is carried out to assess the impact on the environment of extracting and using the metal.

(i) Name three different methods of disposal that might be explored at the final stage of the analysis. **[3]**

(ii) Compare and evaluate the environmental impact of the different methods of disposal identified in part (i). **[4]**

3. Class Y has been asked to carry out some tests on two different solutions: A and B.

Use the data sheet on page 96 to help you to answer these questions.

(a) Solution A is copper chloride.

(i) Describe what you will see when sodium hydroxide is added to the solution. **[1]**

(ii) Describe what you will see when acidified silver nitrate is added to the solution. **[1]**

(b) Class Y then tests Solution B. Here are the results of the test:

Test	Result
Add dilute sodium hydroxide	Red-brown precipitate forms
Add dilute nitric acid, then silver nitrate	Yellow precipitate forms

What is Solution B? Explain how you came to your conclusion. **[3]**

HT

4 Sodium is an element in Group 1 of the periodic table.

(a) The following statements tell us about the reactivity of sodium. Put ticks (✓) in the boxes next to the **two** statements that are true. **[2]**

Sodium burns with a lilac flame. ☐

Sodium reacts with cold water. ☐

Sodium is less reactive than lithium. ☐

Sodium tarnishes in air to make sodium oxide. ☐

(b) The diagram shows a sodium atom.

The nucleus contains .... and ....

(i) Complete the label on the diagram. **[2]**

(ii) Sodium has 11 electrons. Use crosses (x) to draw the electrons on the diagram. **[1]**

(iii) Write down the electron configuration for sodium. **[1]**

(c) Chlorine is an element in Group 7 of the periodic table. Sodium reacts with chlorine to make sodium chloride.

(i) What safety precautions should you take when working with sodium and chlorine? **[2]**

(ii) Write a balanced symbol equation for the reaction between sodium and chlorine. **[2]**

.... + ... ⟶ .... NaCl

(iii) Add state symbols into your equation. **[1]**

(d) Explain what happens to the sodium and chlorine atoms during the reaction. **[6]**

The quality of written communication will be assessed in your answer to this question.

(e) Potassium reacts with chlorine in a similar way to sodium, but the reaction is faster. Explain why. **[3]**

Module C7 (Further Chemistry)

Further chemistry provides an opportunity to study selected chemistry topics in depth. This module looks at:

- the chemical industry today
- the properties of different organic compounds
- how chemical reactions work
- analytical procedures.

The Chemical Industry

The chemical industry synthesises chemicals on different scales according to their value.

Bulk chemicals are made on a large scale, for example:

- ammonia
- sulfuric acid
- sodium hydroxide
- phosphoric acid.

Fine chemicals are made on a small scale, for example:

- drugs
- food additives
- fragrances.

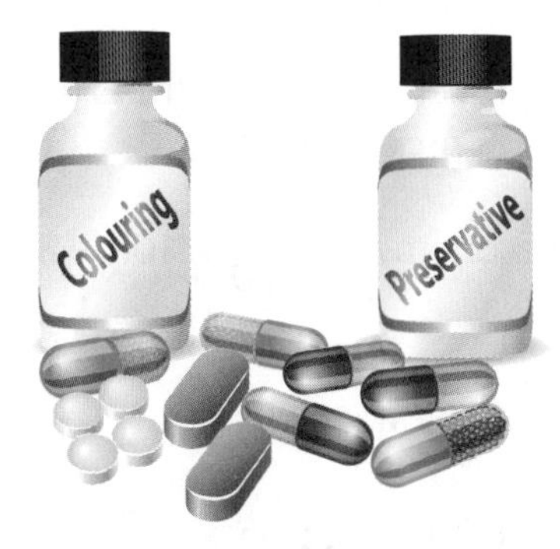

New chemical products or processes are the result of an extensive programme of research and development, for example, researching catalysts for new processes.

Products have to be thoroughly tested to ensure that they are effective and safe to use.

Health and Safety

Governments have a duty to protect people and the environment from any dangers that could occur as a result of procedures involving chemicals.

They impose strict regulations in order to control:

- chemical processes
- the storage of chemicals
- the transportation of chemicals
- the research and development of chemicals.

In the UK, the Health and Safety Executive (HSE) is responsible for the regulation of risks to health and safety arising from the extraction, manufacture and use of chemicals. For example, all hazardous chemicals need to be labelled with standard hazard symbols, such as the ones below:

Corrosive

Harmful

Highly Flammable

Oxidising

Toxic

N.B. The meanings of all the common hazard symbols were given on page 33.

More recently, legislation has been passed to encourage companies to reduce the amount of pollution they produce.

Working in the Chemical Industry

In your exam, you need to be able to interpret information about the work done by people who make chemicals. In general terms, you may need to know that chemists are required to:

- follow standard procedures, for example, when making up a solution (see page 84)
- carry out a titration (see page 58)
- scale up production
- interpret results
- carry out quality assurance.

Green Chemistry

The production of useful chemicals involves several stages, including:

- preparation of feedstocks
- synthesis
- separation of products
- handling of by-products and waste
- monitoring purity.

Green chemistry is based on a number of principles, which if followed lead to more **sustainable** processes, i.e. processes that meet the needs of present generations without compromising future generations.

The principles of green chemistry are:

1 Atom Economy

The final product should aim to contain all the atoms used in the process, thereby reducing waste products and increasing the yield.

$$\text{Atom economy} = \frac{\text{Mass of atoms in the useful product}}{\text{Mass of atoms in the reactants}} \times 100$$

Example

Heating cyclohexanol, $C_6H_{11}OH$, in the presence of a catalyst produces cyclohexene, C_6H_{10}.

$$C_6H_{11}OH \longrightarrow C_6H_{10} + H_2O$$

(a) What is the percentage yield if 10.0g cyclohexanol gives 7.5g of cyclohexene?

Calculate the relative formula masses:

$$(12 \times 6) + (1 \times 12) + 16 = (12 \times 6) + (1 \times 10)$$
$$72 + 12 + 16 = 72 + 10$$
$$100 = 82$$

We know from the equation that 100g of cyclohexanol produces 82g of cyclohexene.

So, theoretical yield from 10g $= \frac{10}{100} \times 82$
$= 8.2\text{g}$

$$\text{Percentage yield} = \frac{\text{Actual yield}}{\text{Theoretical yield}} \times 100$$

Percentage yield $= \frac{7.5}{8.2} \times 100 =$ **91.5%**

N.B. This is a good yield for a preparation, but there is still some waste.

HT You must be able to calculate the masses of reactants and products from balanced equations – see page 57.

(b) Calculate the atom economy, assuming that all the catalyst is re-used.

Total number of atoms in reactant
= 6C, 12H, 1O (RFM = 100)

Total number of **green atoms** in product
= 6C, 10H (RFM = 82)

Total number of **brown atoms** ending as waste
= 2H, 1O (RFM = 18)

Atom economy $= \frac{82}{100} \times 100 =$ **82%**

In this example, the atom economy is relatively high, and the waste atoms form water, so the process is quite green. However, in other reactions, the atom economy may be 50%, and the waste atoms may form pollutants or toxic chemicals. In these examples the reactions would not be green.

2 Use of Renewable Feedstocks

Whenever possible, a renewable raw material should be used. Crude oil (non-renewable) is currently the main source of chemical feedstocks.

Several companies are developing new materials from plants, but plants take up a lot of land. Fertilisers can be used to increase productivity, but they use up a lot of energy during manufacture.

3 Energy Inputs or Outputs

The energy needed to carry out a reaction should be minimised in order to reduce the environmental and economic impact. Where possible, the processes should be carried out at ambient temperature and pressure. Using catalysts makes reactions more efficient and can significantly reduce the amount of energy needed in the process.

4 Health and Safety Risks

Substances used in a chemical process should be chosen to minimise the risk of chemical accidents, including explosions and fires.

Methods need to be developed to detect harmful products before they are made.

5 By-products or Waste

If waste is not made, then it will not have to be cleaned up. Where by-products are made (e.g. carbon dioxide in the burning of fossil fuels), processes should be put in place to deal with them (e.g. carbon dioxide absorbers).

6 Environmental Impact

The environmental impact can be reduced by using alternatives to hazardous chemicals.

Efficient chemical products could be designed that pose minimal harm to people or the environment. They should be able to be broken down into non-toxic substances that do not stay in the environment.

7 Social and Economic Benefits

Social benefits include cleaner air quality and generally less creation of pollution. This will lead to cleaner buildings in towns and improved water quality in rivers and lakes.

Economic benefits include reduced energy costs, as many industrial processes will be operated at lower temperatures and pressures.

Catalysts

The chemical industry carries out research and development to ensure that its processes are **sustainable**. In recent years there has been a lot of research and development into **catalysts**. For example, an exothermic reaction can be started by using only a small amount of energy and, as the catalyst remains unchanged, it can be used over and over again. This makes the process more sustainable. Some industrial processes use enzyme catalysts, which means that the temperature and pH of the reaction must be carefully monitored (see example on page 70).

The **activation energy** is the energy needed to break chemical bonds to start a reaction. Catalysts reduce the activation energy needed for a reaction by providing an alternative route for the reaction. This makes the reaction go faster. This can be illustrated using an energy-level diagram.

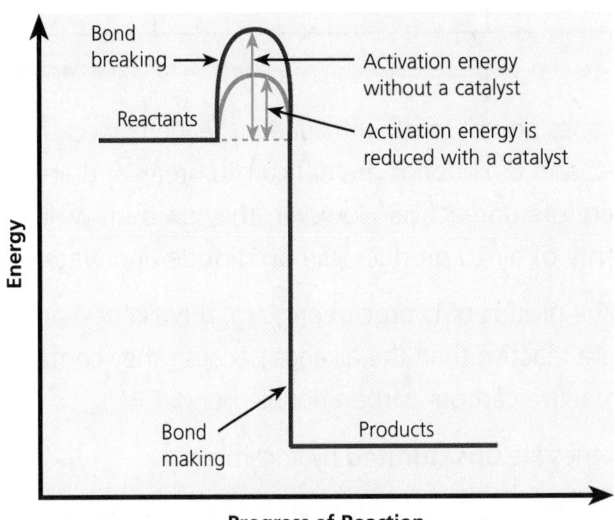

C7 Further Chemistry

Hydrocarbons

Hydrocarbons are made up only of carbon and hydrogen atoms. The 'spine' of a hydrocarbon is made up of a chain of carbon atoms.

There is a group of hydrocarbons called the **alkanes**. In an alkane the carbon atoms are joined together by single carbon–carbon bonds (C–C). So, all the carbon atoms are linked to four carbon or hydrogen atoms by single bonds. This means that all their bonds are single and the hydrocarbon is **saturated**.

The general formula for alkanes is C_nH_{2n+2}

Alkane	Molecular Formula	Ball and Stick Representation	Structural Formula
Methane	CH_4		H | H–C–H | H
Ethane	C_2H_6		H H | | H–C–C–H | | H H
Propane	C_3H_8		H H H | | | H–C–C–C–H | | | H H H
Butane	C_4H_{10}		H H H H | | | | H–C–C–C–C–H | | | | H H H H

Key: ● Carbon atom ○ Hydrogen atom

Alkanes do not react with aqueous reagents because C–C and C–H bonds are difficult to break and are therefore unreactive. However, they do burn well in plenty of air to produce carbon dioxide and water.

Some groups of hydrocarbons, e.g. the alkenes, are more reactive than the alkanes because they contain a reactive carbon–carbon double bond (C=C).

Alkenes are **unsaturated** hydrocarbons.

N.B. You must be able to interpret symbol equations.

HT You must be able to write balanced symbol equations for the combustion of alkanes, for example:

$$CH_4(g) + 2O_2(g) \rightarrow CO_2(g) + 2H_2O(g)$$

$$2C_4H_{10}(g) + 13O_2(g) \rightarrow 8CO_2(g) + 10H_2O(g)$$

Alcohols

Alcohols form a **homologous series** with the **functional group** –OH. The presence of the –OH group gives alcohols their characteristic properties. The general formula for alcohols is $C_nH_{2n+1}OH$, where *n* is the number of carbon atoms.

The two simplest alcohols are **methanol** and **ethanol**:

Methanol, CH_3OH **Ethanol, C_2H_5OH**

```
     H              H   H
     |              |   |
 H — C — O — H  H — C — C — O — H
     |              |   |
     H              H   H
```

Methanol is an important chemical feedstock (i.e. a raw material used for an industrial process.) Methanol can be used in the manufacture of fuels, adhesives, foams, cosmetics and solvents.

Ethanol can be used as a solvent, a fuel and a component in alcoholic drinks.

Physical Properties of Alcohols

Alcohols contain a hydrocarbon chain and an –OH group, so their physical properties can be compared with those of the alkanes and water.

	Boiling Point (°C)	Melting Point (°C)	Density (g/cm^3)	Solubility in Water
Water	100	0	1.0	
Ethanol (alcohol)	78.5	-117	0.789	Soluble
Propanol (alcohol)	97.2	-127	0.803	Soluble
Ethane (alkane)	-88.6	-183	Gas	Insoluble
Propane (alkane)	-42.2	-188	Gas	Insoluble
Octanol (alcohol)	194	-15.5	0.83	Insoluble
Octane (alkane)	126	-56.8	0.7	Insoluble
Decane (alkane)	174.1	-29.7	0.74	Insoluble

From the table, it can be seen that:

- the hydrocarbon chain behaves like the alkane, i.e. it is less dense than water because the long hydrocarbon chains do not mix with water. This is seen with longer chain molecules.
- the –OH group behaves like water, which explains the higher than expected boiling point. The forces between the molecules are stronger than in the alkanes.

Chemical Reactions of Alcohols

Alcohols burn in air to produce carbon dioxide and water. They produce these substances because of the presence of the hydrocarbon chain.

HT The following equation shows what happens when an alcohol burns in air:

$$C_2H_5OH_{(l)} + 3O_{2(g)} \rightarrow 3H_2O_{(g)} + 2CO_{2(g)}$$

Alcohols react with sodium to produce a salt and hydrogen gas. It is the presence of the functional group –OH that allows this reaction to occur, as in the example below. (Sodium ethoxide is a white, solid ionic salt.)

Ethanol + Sodium → Sodium ethoxide + Hydrogen

$$2C_2H_5OH_{(l)} + 2Na_{(s)} \rightarrow 2C_2H_5O^-Na^+_{(s)} + H_{2(g)}$$

Water, alcohols and alkanes react differently with sodium:

- Sodium floats on **water**, melts, rushes around on the surface and rapidly reacts with water giving off hydrogen.
- Sodium sinks in **alcohol**, does not melt and steadily reacts with alcohol giving off hydrogen.
- There is no reaction between sodium and an **alkane**.

The Production of Ethanol

Ethanol can be produced by three methods:

- Synthesis
- Fermentation
- Biotechnology.

Method 1: Synthesis

Raw materials: crude oil or natural gas and steam.
Product: produces up to 96% pure ethanol on an industrial scale and is used as a feedstock, solvent or fuel.

1. **Preparation of Feedstock**
   Crude oil undergoes fractional distillation. The fractions containing the long-chained hydrocarbons are collected. The alkanes are then heated until they vaporise. The molecules are cracked by passing the vapour over a catalyst at:
   - high temperature (300°C)
   - 60–70 atmospheres pressure.

   After purification by further fractional distillation, the ethene molecules produced in the cracking process are used for feedstock.

   The remaining 4% water is removed by zeolites, which absorb the water molecules to produce pure ethanol. This method replaced the old dehydration method, which used more energy and produced carcinogenic by-products.

2. **Synthesis of Ethanol**
   Ethene is continuously reacted with steam at a moderately high temperature and pressure by passing the gases over a catalyst (phosphoric acid).

Ethene + Steam → Ethanol
$C_2H_4(g) + H_2O(g) \rightarrow C_2H_5OH(g)$

3. **Recycling**
   Any unreacted products are recycled and fed through the system again.

Ethene may also be obtained by cracking ethane obtained from natural gas:

$$C_2H_6(g) \rightarrow C_2H_4(g) + H_2(g)$$

Method 2: Fermentation

Raw materials: natural sugars, yeast and water.
Product: produces up to approximately 15% alcohol by volume.

Water and yeast are mixed with sugars at just above room temperature. **Enzymes** (biological catalysts), found in the yeast, react with the sugars to form ethanol and carbon dioxide. The carbon dioxide is allowed to escape from the reaction vessel, but air is prevented from entering it.

Water + Sugars + Yeast → Ethanol + Carbon dioxide
$C_6H_{12}O_6(aq) \rightarrow 2C_2H_5OH(aq) + 2CO_2(g)$

Temperature and **pH** are important factors to consider when determining optimum conditions for the fermentation process. Enzymes use a 'lock and key' mechanism, which means that a specific reactant fits into a specific enzyme. If the temperature of the reaction rises too much, the enzyme is **denatured** (the shape is irreversibly changed) and the reactant can no longer fit into the enzyme.

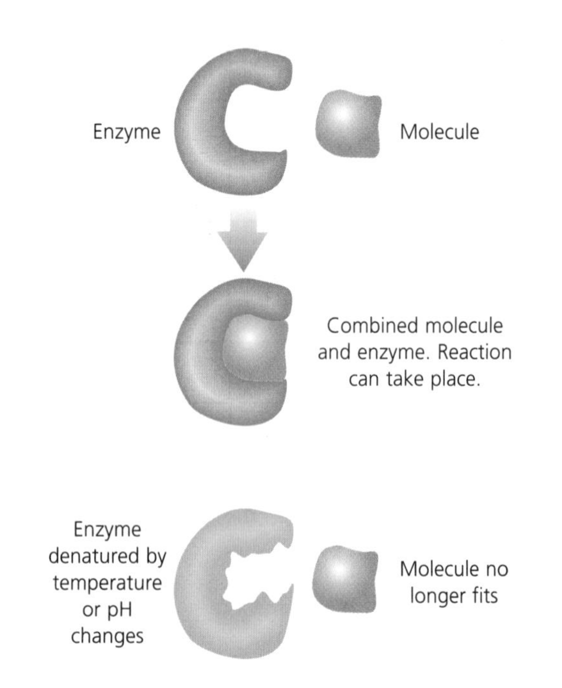

If the pH of the mixture changes too much, the enzyme may also become denatured due to attractions of excess H^+ or OH^- ions.

When ethanol solution is manufactured by fermentation, the concentration is limited. The main limiting factors are:

- the amount of sugar in the mixture (the reaction will stop once the sugar is all used up)
- the enzymes in the yeast (if the concentration of ethanol rises above approximately 15%, the enzymes will die of alcohol poisoning and the reaction will stop).

When the fermentation reaction is over, the concentration of the ethanol may be **increased** by distilling the mixture. This distillation process is used to produce spirits such as whisky and brandy. In some distilleries the whisky is distilled twice.

Method 3: Biotechnology

Raw materials: waste biomass and genetically engineered *E.coli* bacteria.
Product: produces up to 95% pure alcohol.

The biotechnology method uses genetically modified *E.coli* bacteria that have had new genes introduced. The new genes allow the bacteria to digest all the sugars in the biomass and convert them into ethanol.

This means that a wider range of biomass, such as wood waste, corn stalks and rice hulls, can be converted to ethanol rather than remaining as waste. This method still needs a variety of organic substances for bacterial metabolism and growth.

The optimum temperature for this reaction is 25–37°C. The optimum pH level needs to remain fairly constant otherwise the enzyme will be **denatured**.

Interpreting the Processes

In your exam you will need to be able to interpret information about the three processes used to produce ethanol and evaluate their sustainability.

For example, you may be supplied with the following information and be asked a question, such as 'Is fermentation or biotechnology the most sustainable method of producing ethanol? Explain your answer'.

Ethanol is used as a fuel, a solvent in paints and cosmetics and as a starting material for making polymers. Ethanol can be produced by three different methods; the advantages and disadvantages are listed below:

Synthetic Method

- High energy usage.
- Toxic by-products produced.
- A purification stage is needed.
- Uses non-renewable raw materials.
- Ethene is converted to ethanol using steam and a catalyst.

Fermentation Method

- Uses renewable raw materials, e.g. sugar cane.
- Carbon-neutral process.
- Some household waste can be used to produce ethanol.
- Large areas of land are needed to grow specific crops as the raw materials.
- Only part of the plant material is used; the rest can be used to make animal feedstocks.
- Sugars are converted into ethanol and carbon dioxide.

Biotechnology Method

- Raw materials are waste biomass and genetically engineered *E. coli* bacteria not specifically grown for ethanol production.
- Genetically engineered *E. coli* bacteria are used to convert plant sugars into ethanol.
- Carbon-neutral process.

C7 Further Chemistry

Carboxylic Acids

Carboxylic acids form a **homologous series** with the **functional group** –COOH. The presence of the –COOH group gives carboxylic acids their characteristic properties. The two simplest carboxylic acids are **methanoic acid** and **ethanoic acid**:

Methanoic Acid, HCO_2H

```
       O
      //
H – C
      \
       O – H
```

Ethanoic Acid, CH_3CO_2H

```
    H     O
    |    //
H – C – C
    |    \
    H     O – H
```

Vinegar is a dilute solution of ethanoic acid.

Carboxylic acids are found in many substances, and some have unpleasant smells and tastes. For example, they are responsible for:

- the aroma of a sweaty training shoe
- the taste of rancid butter.

The general formula of carboxylic acids is $CH_{2n+1}COOH$.

Chemical Reactions of Carboxylic Acids

Carboxylic acids are weak acids. In solution they have a pH from 3 to 6. Like all acids, they can react with metals, alkalis and carbonates to produce carboxylic acid salts, but they are less reactive than strong acids, e.g. hydrochloric acid. Some examples are shown below.

Reaction of a carboxylic acid with a **metal**:

Ethanoic acid + Sodium ➔ Sodium ethanoate + Hydrogen

Reaction of a carboxylic acid with an **alkali**:

Ethanoic acid + Sodium hydroxide ➔ Sodium ethanoate + Water

Reaction of a carboxylic acid with a **carbonate**:

Ethanoic acid + Sodium carbonate ➔ Sodium ethanoate + Water + Carbon dioxide

Fats

Fats and oils are naturally occurring esters (see below). Living organisms make them to use as an energy store. Fats are the esters of:

- **glycerol**, an alcohol with three –OH groups
- **fatty acids**, which are carboxylic acids with very long hydrocarbon chains.

Animal fats, such as lard and fatty meat, are mostly **saturated molecules**. This means they have single carbon–carbon bonds, and the molecules are unreactive. For example, stearic acid is a saturated fat.

Stearic Acid

Only single carbon–carbon bonds present.

$$CH_3-(CH_2)_{15}-CH_2-COOH$$

Vegetable oils, such as olive oil and sunflower oil, are mostly **unsaturated molecules**. This means that they contain some double carbon–carbon bonds (C=C). The presence of the C=C bonds means that the molecules are reactive. For example, vegetable oils (hydrogenated fats) are unsaturated fats.

Unsaturated Fat

Double carbon–carbon bond

```
CH3–(CH2)7       (CH2)7 – COOH
          \     /
           C = C
           |   |
           H   H
```

Esters

Carboxylic acids react with alcohols to form **esters**, as in the following example:

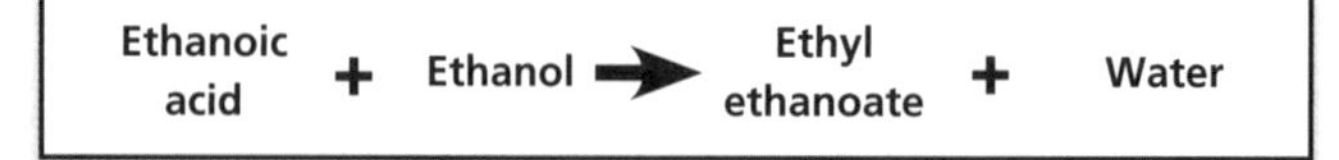

This reaction is carried out in the presence of a **catalyst**, i.e. concentrated sulfuric acid. Esters have distinctive smells that are responsible for the smells and flavours of fruits. Due to their sweet smell, they are often used in the manufacture of perfumes, fragrances and food products (for artificial flavours such as raspberry, pear and cherry). Esters are also found in products such as solvents in **adhesives** and **plasticisers** because they contain hydrocarbon chains.

Preparing Esters

An ester can be prepared by the following method:

1. Ethanol and excess ethanoic acid are **heated under reflux** in the presence of concentrated sulfuric acid.

Water out
Condenser
Water in
Reactants (ethanol and ethanoic acid) and catalyst (concentrated sulfuric acid)
Round-bottom flask
Heating mantle

2. The ester is removed by **distillation**. (Ethyl ethanoate boils at 77°C.)

Distillation Equipment

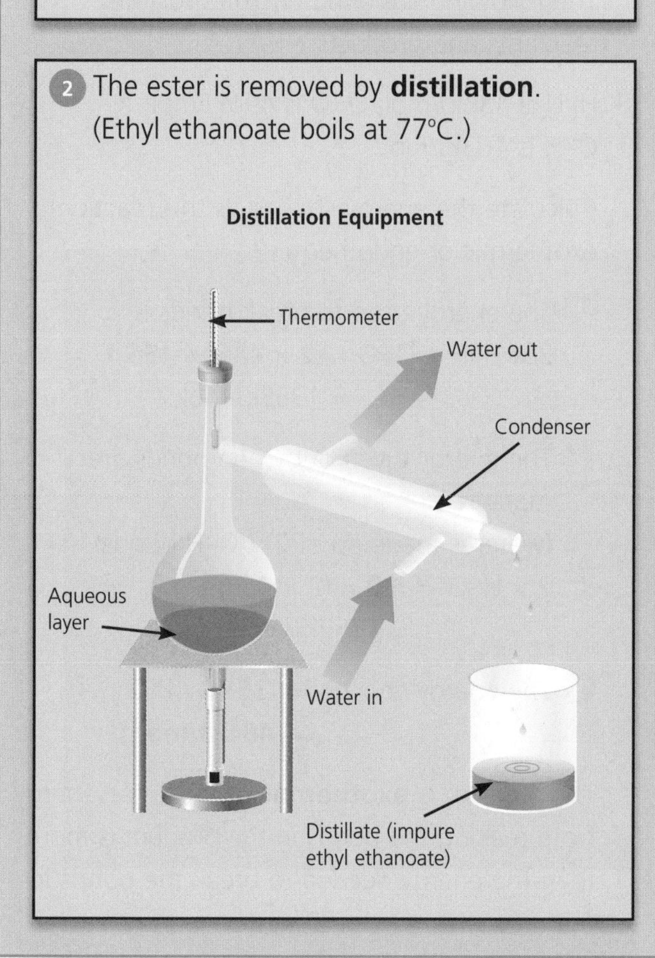

3. The distillate is transferred to a separating funnel, where it is purified. A solution of sodium carbonate is added, and the mixture is shaken up. This mixture will react with any remaining acid and extract it into the aqueous phase. The aqueous phase is then run off leaving the ester in the funnel.

Glass stopper
Impure ester (distillate)
Aqueous phase (sodium carbonate solution)

4. The product is transferred to a conical flask and anhydrous calcium chloride is added to remove any remaining water molecules. The calcium chloride is removed later by filtration.

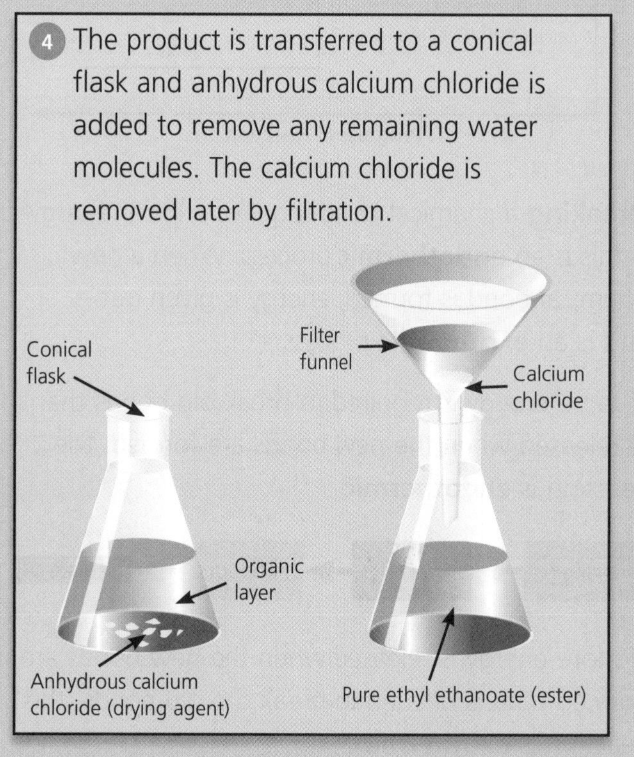

N.B. You must be able to explain what is meant by 'heating under reflux', 'distillation', 'purification', 'using a tap funnel' and 'drying'.

Energy Changes

In Module C6 we covered exothermic and endothermic reactions, and energy-level diagrams. It is important that you remind yourself of these ideas before going on to learn more about the energy changes that occur during chemical reactions.

Making and Breaking Bonds

In a chemical reaction, new substances are produced. In order for this to happen, the bonds in the reactants must be broken and new bonds made to form the products.

The **activation energy** is the energy needed to start a reaction, i.e. to break old bonds. This can be shown on an energy-level diagram.

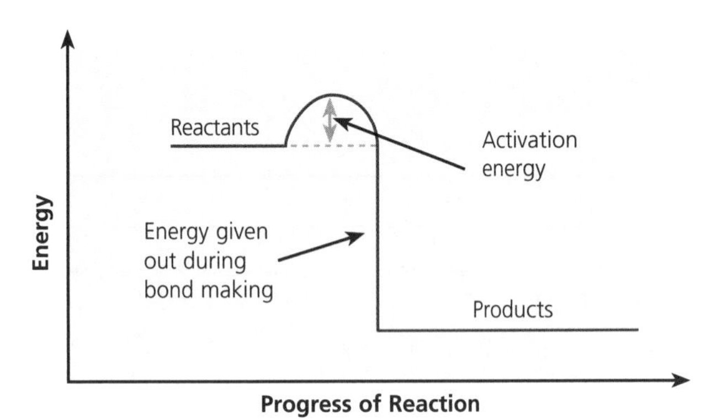

Breaking a chemical bond requires a lot of energy – this is an **endothermic** process. When a new chemical bond is formed, energy is given out – this is an **exothermic** process.

If more energy is required to break old bonds than is released when the new bonds are formed, the reaction is **endothermic**.

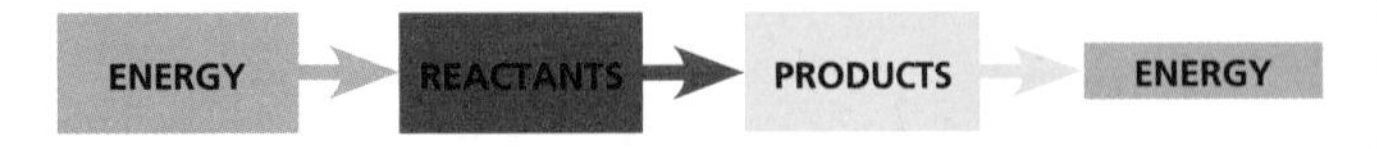

If more energy is released when the new bonds are formed than is needed to break the old bonds, the reaction is **exothermic**.

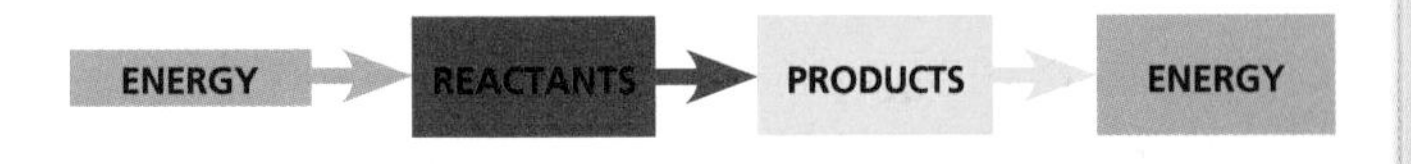

Energy Calculations and Supplied Bond Energies

HT

Example 1

Hydrogen is burned in oxygen to produce steam:

Hydrogen + Oxygen → Steam
$2H_2(g) + O_2(g) \rightarrow 2H_2O(g)$

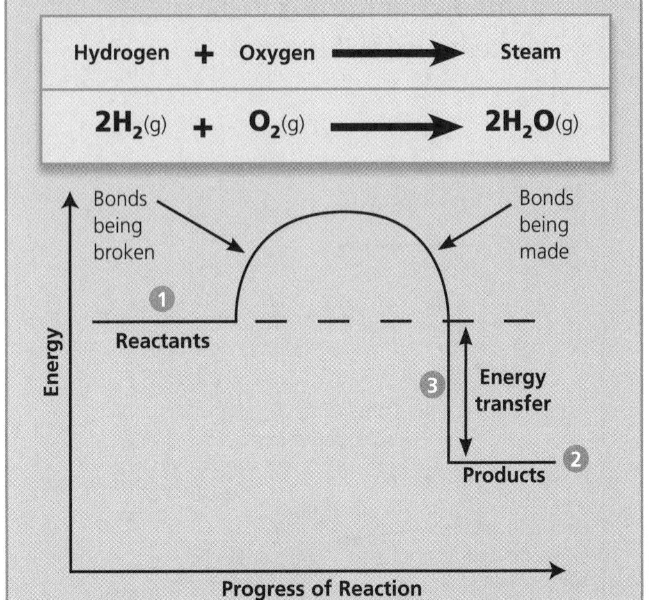

The following are bond energies for the reactants and products:

H–H is 436kJ/mol; O=O is 496kJ/mol;
O–H is 463kJ/mol

Calculate the energy change. Is this reaction exothermic or endothermic?

1. The energy used to break bonds is:
   2 × H–H + O=O = (2 × 436) + 496
   = 1368kJ/mol
2. The energy given out when bonds are made is:
   (water is made up of 2 × O–H bonds)
   2 × H–O–H = 2 × (2 × 463)
   = 1852kJ/mol
3. The energy change = 1368 – 1852
   = **-484kJ/mol**

The reaction is **exothermic** because the energy from making the bonds in the product is more than the energy needed to break the bonds in the reactants.

HT **Example 2**

Hydrogen and halogens react together to form hydrogen halides. For example, the formation of hydrogen chloride is shown below.

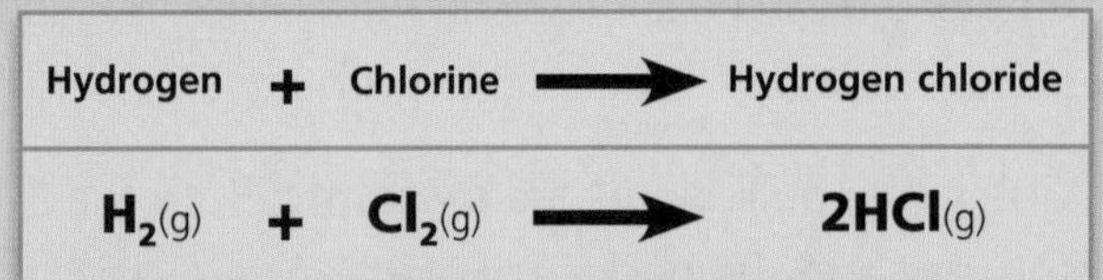

Hydrogen + Chlorine → Hydrogen chloride
$H_2(g) + Cl_2(g) \longrightarrow 2HCl(g)$

The following are bond energies for the reactants and products:

H–H is 436kJ/mol; Cl–Cl is 243kJ/mol; H–Cl is 432kJ/mol.

Calculate the energy changes. Is this reaction exothermic or endothermic?

1. The energy used to break bonds is:
   H–H + Cl–Cl = 436 + 243
   = 679kJ/mol
2. The energy given out when bonds are made is:
   2 × H–Cl = 2 × 432
   = 864kJ/mol
3. The energy change is = 679 – 864
   = **-185kJ/mol**

The reaction is **exothermic**.

The bond energy calculations help increase our confidence in explaining the energy changes that take place in chemical reactions. This is because they agree with the observations that scientists have made, i.e. in exothermic reactions the temperature goes up and energy is given out.

Research scientists find this data very useful. For example, when looking for non-polluting fuels, scientists need to know how much energy is given out during the reaction. Example 1 shows that burning hydrogen in oxygen is a very exothermic reaction that only produces steam. Once the activation energy has been overcome, and the reaction is started, there is enough energy to sustain the reaction. This is why hydrogen-fuelled cars may be used more in the future.

Reversible Reactions

Some chemical reactions are reversible, i.e. the products can react together to produce the original reactants. In an equation, $\rightleftharpoons$ shows that the reaction can go in either the forward or reverse direction, depending on the conditions:

$$A + B \rightleftharpoons C + D$$

This means that:

- A and B can react together to produce C and D
- C and D can react together to produce A and B.

For example, solid ammonium chloride decomposes when heated to produce ammonia and hydrogen chloride gas, both of which are colourless. Hydrogen chloride gas and ammonia react to produce white clouds of ammonium chloride.

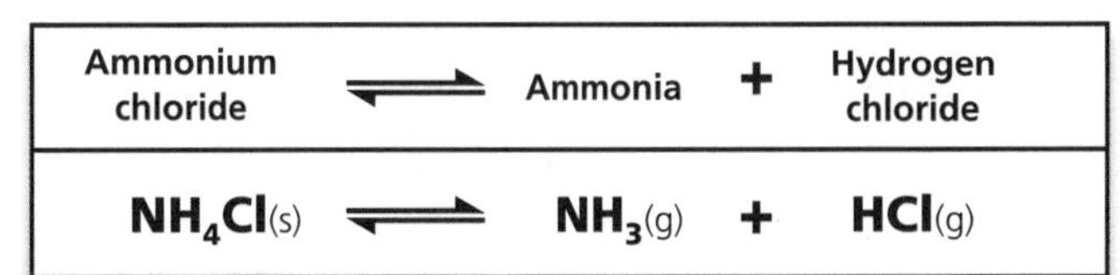

Ammonium chloride ⇌ Ammonia + Hydrogen chloride
$NH_4Cl(s) \rightleftharpoons NH_3(g) + HCl(g)$

Reversible Reaction of Ammonium Chloride

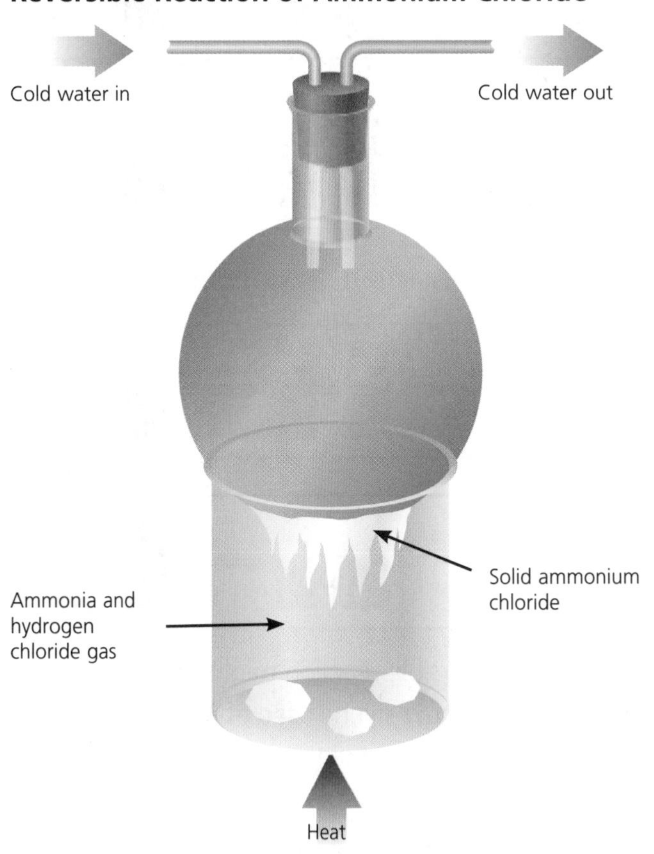

Equilibrium

A reversible reaction will reach a state of **equilibrium** if it is in a **closed system** (a system where no reactants are added and no products are taken away).

At equilibrium the reaction appears to have stopped. However, neither the forward reaction (from left to right) nor the backward reaction (from right to left) are complete as both reactants and products are present at the same time. The concentration of the reactants and products does not change.

The relative amounts of all the reacting substances at equilibrium depend on the conditions of the reaction. For example, the following diagram represents a reaction:

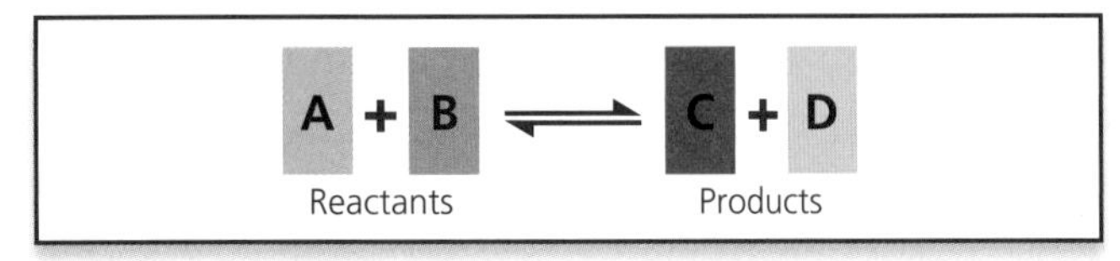

If the **forward reaction** (the reaction that produces the products C and D) is **endothermic** then:

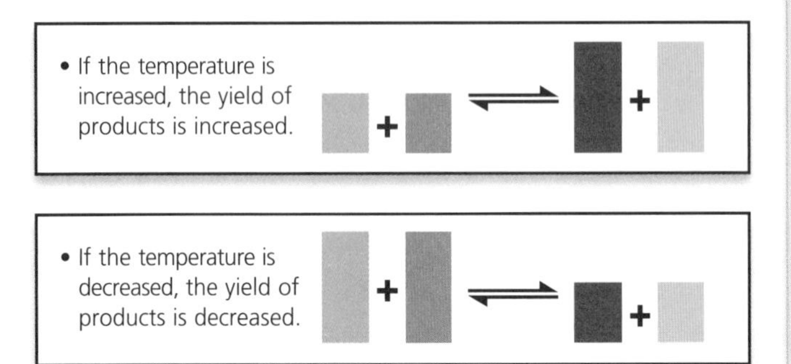

If the **forward reaction** is **exothermic** then:

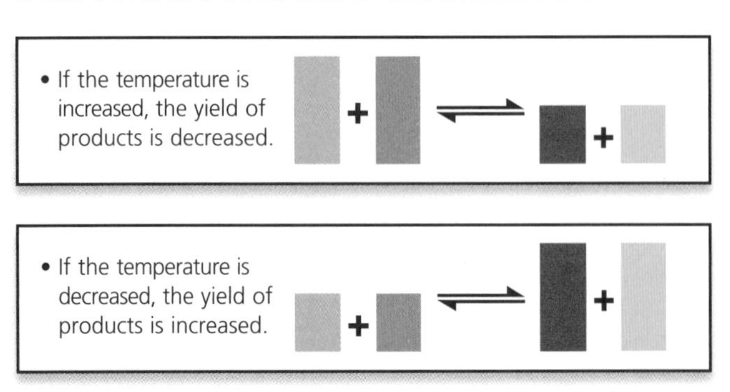

Although a reversible reaction might not go to completion, it could still be used efficiently in an industrial process, e.g. the Haber process for the industrial manufacture of ammonia (see page 77).

HT Achieved Equilibrium

Once equilibrium is achieved, the concentration of the reactants and products does not change. The equilibrium can be approached from either direction, i.e. the 'reactant' side or the 'product' side.

Chemical equilibriums are **dynamic**. Both the forward and the backward reactions are still occurring, but at the same rate. Therefore, there is no overall change in concentration of the substances.

The diagrams below show what happens to iodine particles when you shake a solution of iodine in an organic solvent with aqueous potassium iodide.

At the Start

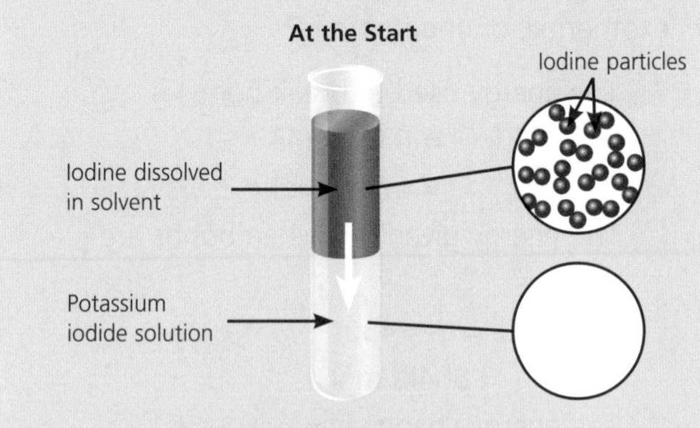

Approaching Equilibrium
(Particles move between the solvent and solution)

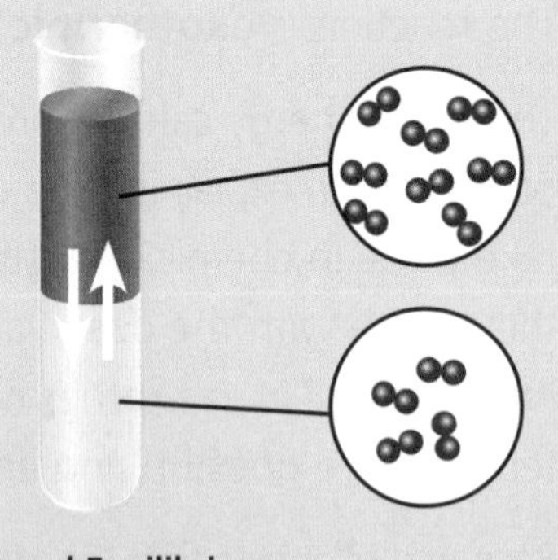

Achieved Equilibrium
(Particles continue to move in both directions at the same rate, so equilibrium is maintained. This is called dynamic equilibrium.)

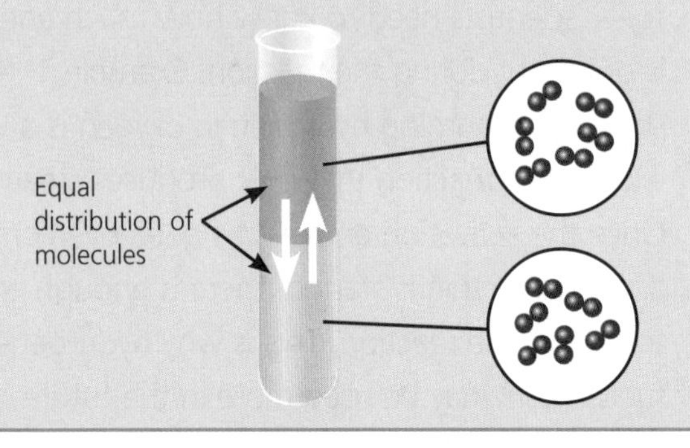

Ammonia is a Very Important Chemical

Ammonia in Society and its Effect on the Environment

Ammonia is a very important chemical as it is needed to make fertilisers (ammonium nitrate), nitric acid and explosives. Traditionally ammonia was made using sources of nitrate, such as guano (bird droppings).

The Haber process was a major scientific breakthrough as it allowed nitrogen from the air to be converted into ammonia on an industrial scale. This in turn has affected both society and the environment. For example:

- the ability to manufacture explosives prolonged the First World War
- a good supply of fertilisers has allowed the increasing world population to be fed and more land is needed to build houses for the growing population
- nitrates have polluted rivers.

The Haber Process

The raw materials for the manufacture of ammonia by the Haber process are nitrogen and hydrogen. Nitrogen is extracted from the air by fractional distillation of liquid air. Hydrogen is obtained by reacting natural gas (methane) with steam:

$$CH_4(g) + H_2O(g) \longrightarrow 3H_2(g) + CO(g)$$

The purified nitrogen and hydrogen are passed over an iron catalyst at a temperature of about 450°C and a high pressure of about 200 atmospheres. The reaction is reversible and exothermic:

Nitrogen	+	Hydrogen	$\rightleftharpoons$	Ammonia
$N_2(g)$	+	$3H_2(g)$	$\rightleftharpoons$	$2NH_3(g)$

Some of the hydrogen and nitrogen reacts to form ammonia. Some of the ammonia produced decomposes into nitrogen and hydrogen. However, the gases do not stay in the reaction vessel long enough to reach equilibrium. To increase the overall yield, the unreacted nitrogen and hydrogen are **recycled**. The conditions for the Haber process are as follows:

- **Low Temperature**
  The reaction in the forward direction is exothermic. If the temperature is lowered, the equilibrium will shift to the right to increase the temperature. The amount of ammonia formed will increase. Lowering the temperature slows down the reaction and a catalyst of iron is used to speed up the reaction.
- **High Pressure**
  There are four molecules of gas on the left-hand side and two molecules of gas on the right-hand side. If the pressure is increased, the equilibrium will shift to the right to lower the pressure. The amount of ammonia formed will increase.

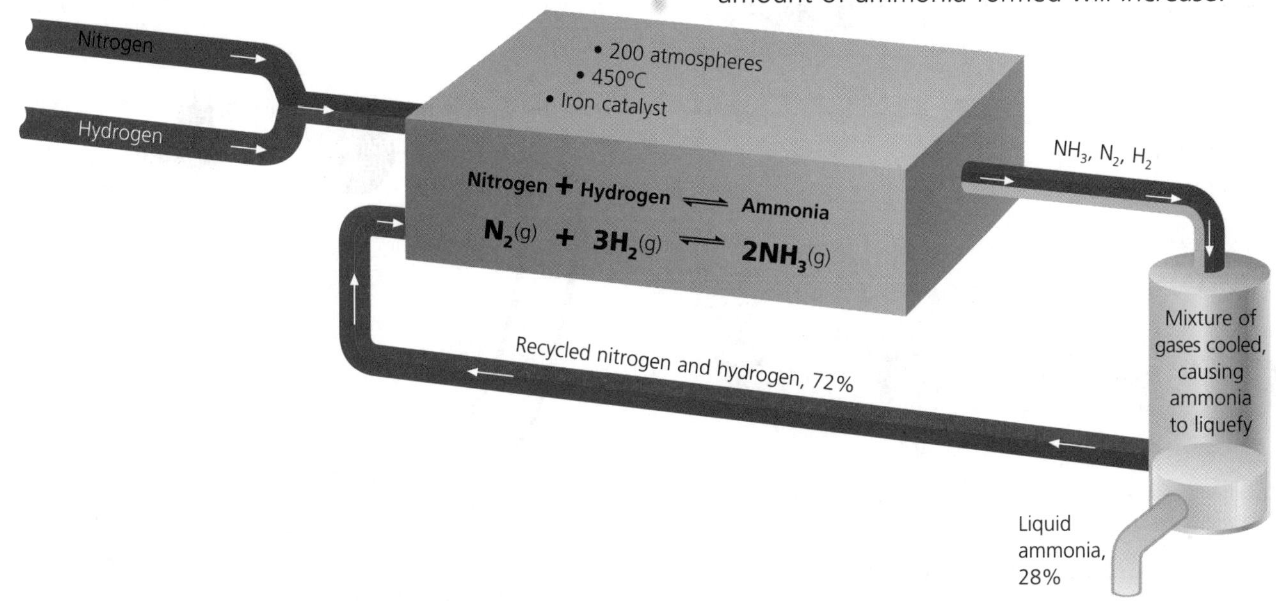

HT A Compromise Solution

Altering the temperature and pressure can have a big impact on the production of ammonia in the Haber process. The conditions have to be chosen very carefully to be economically viable and to make sure they can meet demand.

The formation of ammonia is exothermic so a low temperature increases the yield, but the reaction is very slow. A high temperature makes the reaction faster but produces a lower yield. So a compromise is reached.

The volume of ammonia produced is less than the total volume of the reactants (nitrogen and hydrogen) so a high pressure favours the production of ammonia, but this is very expensive. A low pressure is more affordable, but this produces a low yield. So, yet again, a compromise is reached.

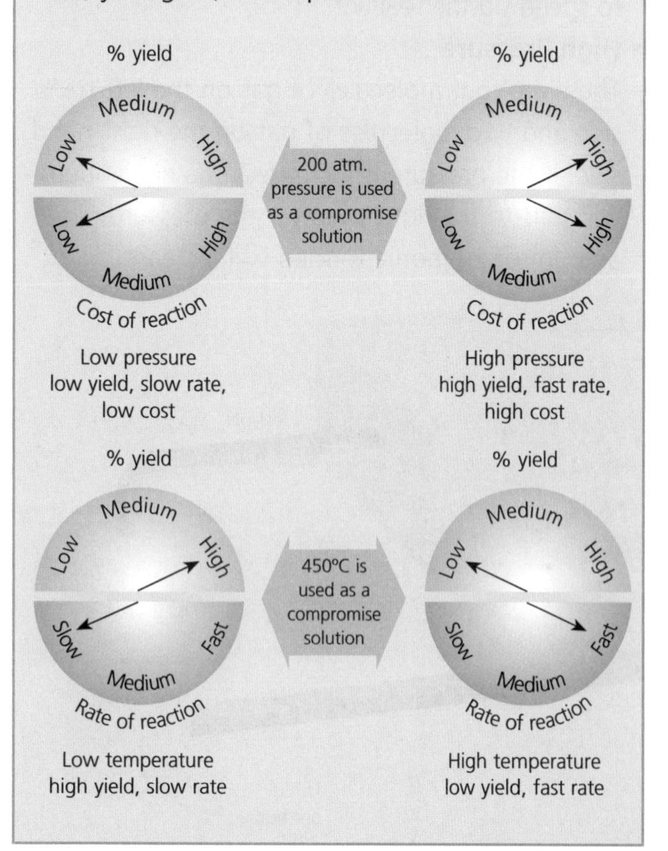

The Nitrogen Cycle

The **nitrogen cycle** shows how nitrogen and its compounds are recycled in nature. Nitrogen is a vital element of all living things and is used to make proteins, which are used in plant and animal growth. All enzymes are proteins. Nitrogen gas in the air cannot be used by plants and animals as it is inert (unreactive). Plants can only use it in the form of nitrates. The main processes in the nitrogen cycle are as follows:

1. **Nitrogen-fixing bacteria** convert atmospheric nitrogen into nitrates in the soil (see page 79).
2. When plants are eaten the nitrogen becomes animal protein.
3. Dead organisms and waste contain ammonium compounds.
4. **Decomposers** convert urea, faeces and protein from dead organisms into ammonium compounds.
5. **Nitrifying bacteria** convert ammonium compounds into nitrates in the soil.
6. **Denitrifying bacteria** convert nitrates into atmospheric nitrogen and ammonium compounds.

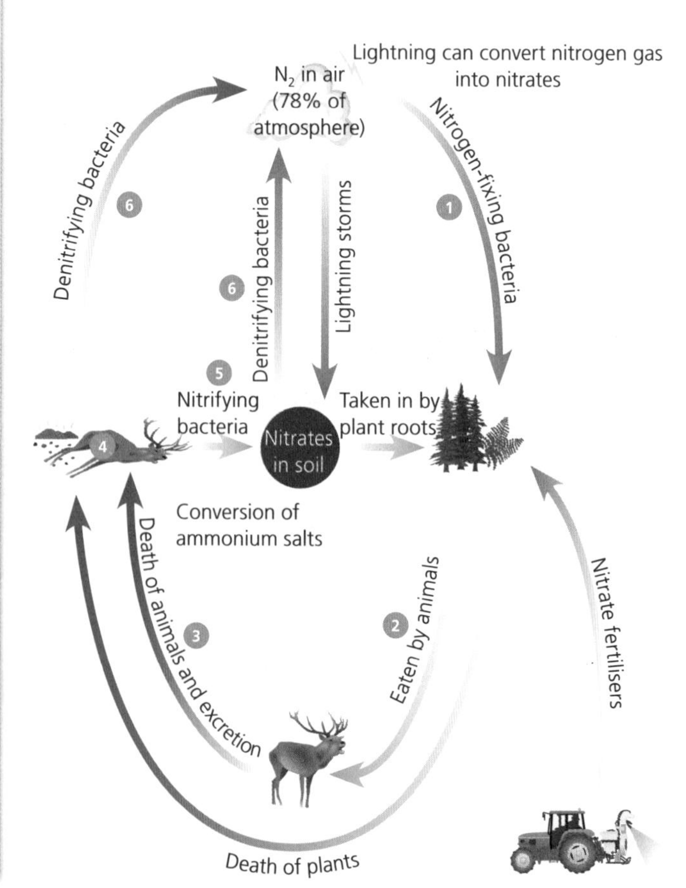

Nitrogen Fixation

Nitrogen fixation is the process by which nitrogen gas in the atmosphere is converted into more useful nitrogen compounds, such as ammonia, nitrates and nitrogen oxides. This can be done by nitrogen-fixing bacteria, which contain the enzyme **nitrogenase**. Nitrogenase is a biological catalyst that allows the process of fixing nitrogen to take place at room temperature and pressure.

Some nitrogen-fixing bacteria live in the soil, while others are found in the **root nodules** of legumes such as peas, beans and clovers.

Nitrogen can also be fixed during **lightning storms**. The enormous energy of lightning breaks the strong covalent bonds in the nitrogen molecule. This enables the nitrogen atoms to react with oxygen in the air, forming nitrogen oxides. Nitrates are formed as the nitrogen oxides dissolve in rainwater.

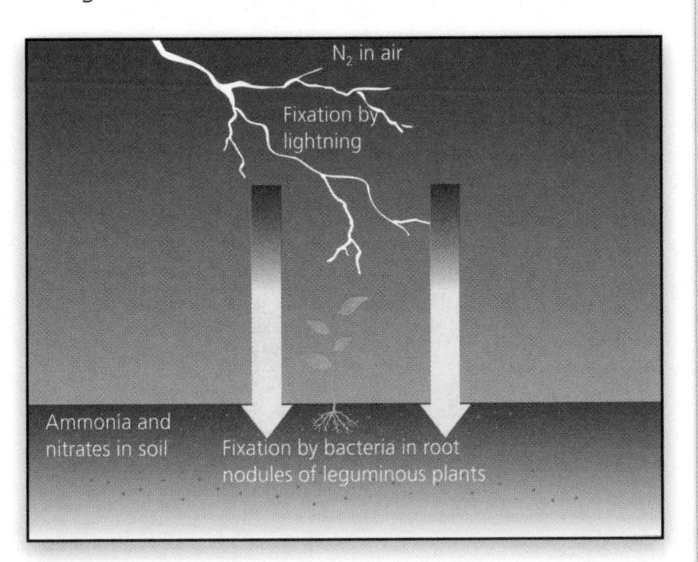

Industrial Methods of Nitrogen Fixation

Nitrogen can be fixed on an industrial scale through chemical reactions such as the Haber process, in which nitrogen from the air is combined directly with hydrogen to form ammonia (see page 77). This method of production is not very sustainable:

- It is expensive to run as high temperatures and pressures are involved, requiring specialised equipment.
- Hydrogen comes from natural gas, which is non-renewable.
- It uses a lot of energy.
- The yield is low – only about 15%.

Looking to the Future

Scientists are working hard to develop new catalysts to try to improve the efficiency of the Haber process. They are particularly interested in producing a new catalyst that mimics the natural enzyme, nitrogenase. They know that it contains clusters of iron (Fe), molybdenum (Mo) and sulfur (S), and have been successful in making a molecule that shows some catalytic activity.

If successful, this would be a major breakthrough as it would enable nitrogen to be fixed on an industrial scale at room temperature and pressure. It would provide a very sustainable source of ammonia and would solve most of the issues listed above.

Over the years there have been some very specific developments in the industrial methods of nitrogen fixation, many of them involving more efficient catalysts. In the future scientists may need to look for a new source or feedstock for hydrogen if reserves of natural gas become depleted. Hydrogen obtained by the electrolysis of water could be an option, but this would depend on the availability of a large renewable energy source, e.g. solar energy or hydropower.

Evaluating Sustainability

When presented with information about the different nitrogen fixation processes, you must be able to interpret data to evaluate the sustainability of each process.

To help you do this, you will need to remind yourself of the principles of green chemistry (see pages 66–67). It will also be helpful to look at the examples given on the production of ethanol (see pages 69–71) to remind yourself of the key pieces of information when evaluating the sustainability of processes.

Analysis

There are two types of analytical procedure:

- **Qualitative** methods
- **Quantitative** methods.

Qualitative analysis is any method used to identify the **chemicals** in a substance, e.g. using an indicator to find out if acids are present or using thin layer chromatography (see page 81).

Quantitative analysis is any method used to determine the **amount** of chemical in a substance, e.g. carrying out an acid–base titration to find out how much acid is present.

Many of the analytical methods you have learned are based on samples in solutions.

There are **standard procedures** for the collection, storage and preparation of samples for analysis.

When collecting data, it is very important that the samples are representative of the **bulk** of the material under test.

This is achieved by collecting multiple samples at random. After a sample has been collected, it should be stored in a sterile container to prevent change or deterioration.

The container should be sealed, labelled and stored in a safe place.

Using a system of common practices and procedures, such as ensuring that samples are not contaminated, can increase reliability since there is less room for human error. Different people can also repeat a test on the same sample and produce the same result.

Chromatography

Chromatography is a technique used to find out what unknown mixtures are made up of.

Paper Chromatography

1. If the substance to be analysed is a solid, dissolve it in a suitable solvent (the solvent used will depend on the solubility of the substance).
2. Place a spot of the resulting solution onto a sheet of chromatography paper on the pencil line and allow it to dry.
3. Place the bottom edge of the paper into a suitable solvent.
4. The solvent rises up the paper, dissolving the 'spot' and carrying it, in solution, up the paper.
5. The different chemicals in the mixture become separated because their molecules have different sizes and properties. The molecules that bind strongly to the paper travel a shorter distance than the molecules that bind weakly to the paper.

The chromatogram can then be compared to standard chromatograms (**standard reference materials**) of known substances to identify the different chemicals.

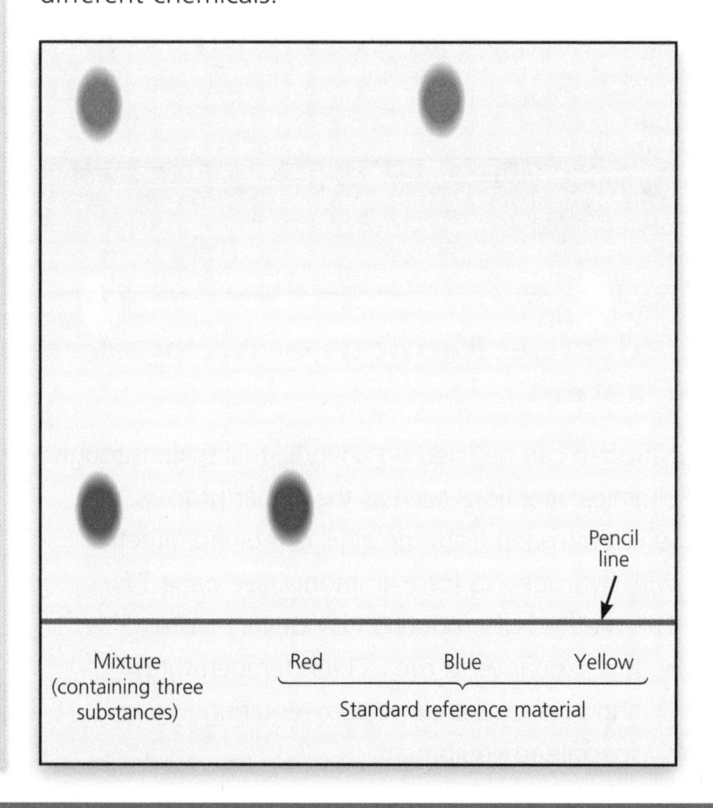

The solvent that is used to move the solution is called the **mobile phase**.

A range of aqueous and non-aqueous solvents may be used. Aqueous solvents are water based, whereas non-aqueous solvents are made from organic liquids such as alcohols.

The medium that it moves through is called the **stationary phase**. In this case the paper is the stationary phase.

A chromatogram is formed when the chemicals come out of solution and bind to the paper, i.e. they move between the mobile phase and the stationary phase.

For each component of the sample, a dynamic equilibrium is set up between the stationary and mobile phase.

Different molecules in the sample mixture travel different distances according to how strongly they are attracted to the molecules in the stationary phase, in relation to their attraction to the solvent molecules.

Therefore, the overall separation depends on the distribution of the compounds in the sample between the mobile and stationary phases.

Thin Layer Chromatography

Thin layer chromatography (TLC) is similar to paper chromatography. However, the stationary phase is a thin layer of adsorbent material (e.g. silica gel, alumina or cellulose) supported on a flat, unreactive surface (e.g. a glass, metal or plastic plate).

There are several advantages of thin layer chromatography over paper chromatography. The advantages include:

- faster runs
- more even movement of the mobile phase through the stationary phase
- a choice of different adsorbents for the stationary phase (which can increase the attraction between molecules in the mixture and the stationary phase).

As a result, thin layer chromatography usually produces better separations for a wider range of substances.

Locating Agents

Some chromatograms have to be developed using **locating agents** to show the presence of colourless substances:

- Colourless spots can sometimes be viewed under ultraviolet (UV) light and then marked on the plate.
- The chromatogram can be viewed by being sprayed with a chemical that reacts with the spots to cause coloration.

Chromatography

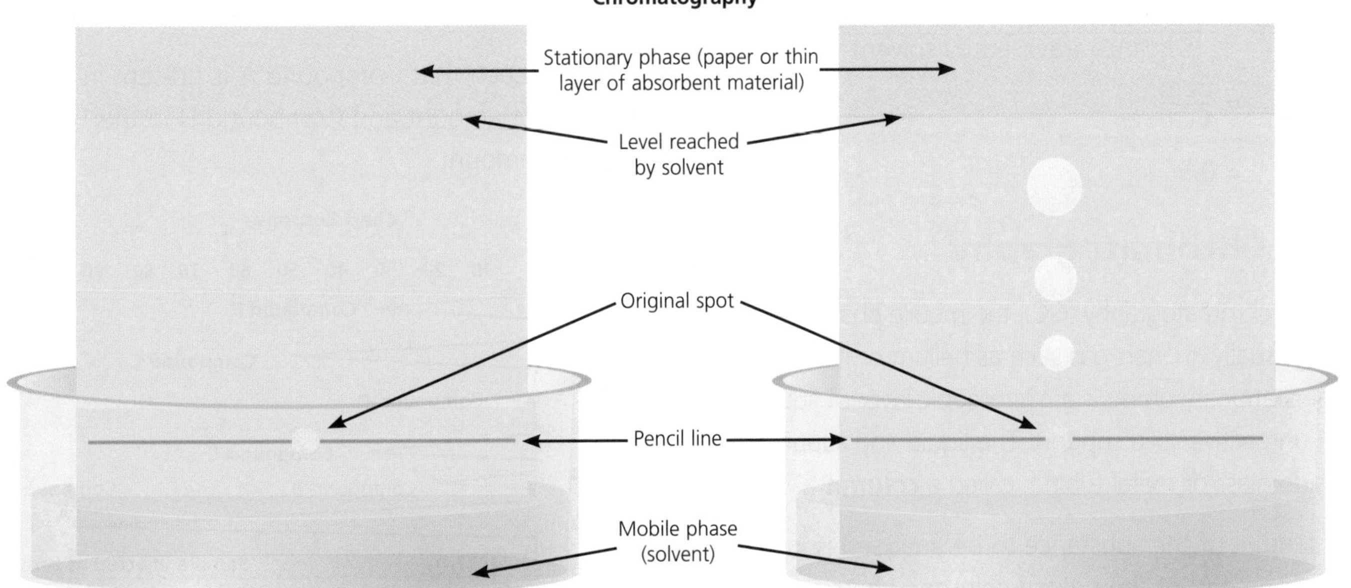

R_f Value

In paper and thin layer chromatography, the movement of a substance relative to the movement of the solvent front is known as the **R_f value**:

$$R_f \text{ value} = \frac{\text{Distance travelled by substance}}{\text{Distance travelled by solvent}}$$

Calculating the R_f value can aid in the identification of unknown substances.

Example

The diagram below shows the distance travelled by a substance and the distance travelled by the solvent. Calculate the R_f value.

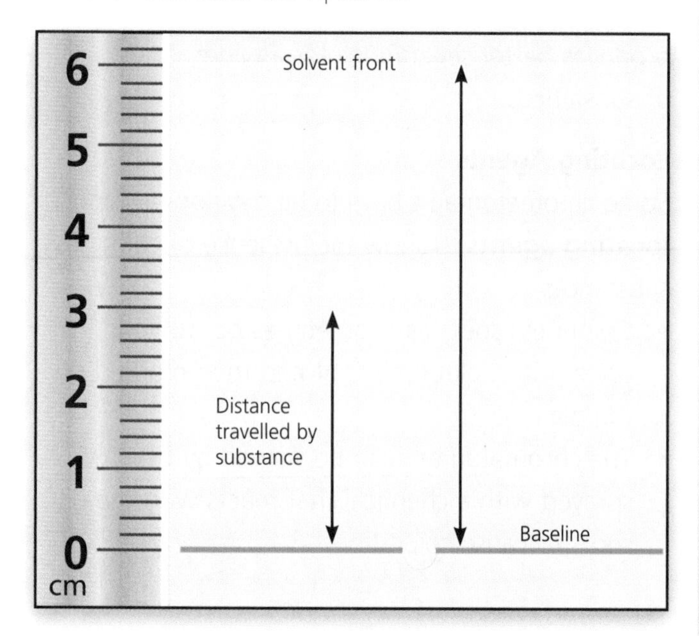

$$R_f \text{ value} = \frac{\text{Distance travelled by substance}}{\text{Distance travelled by solvent}}$$

$$= \frac{3\text{cm}}{6\text{cm}}$$

$$= \mathbf{0.5}$$

Gas Chromatography

In gas chromatography (GC), the mobile phase is a carrier gas, usually an inert gas such as helium or nitrogen. The stationary phase is a microscopic layer of liquid on an unreactive solid support. The liquid and support are inside glass or metal tubing, called a **column**.

A sample of the substance to be analysed is injected into one end of the heated column, where it vaporises. The carrier gas then carries it up the column, where separation takes place.

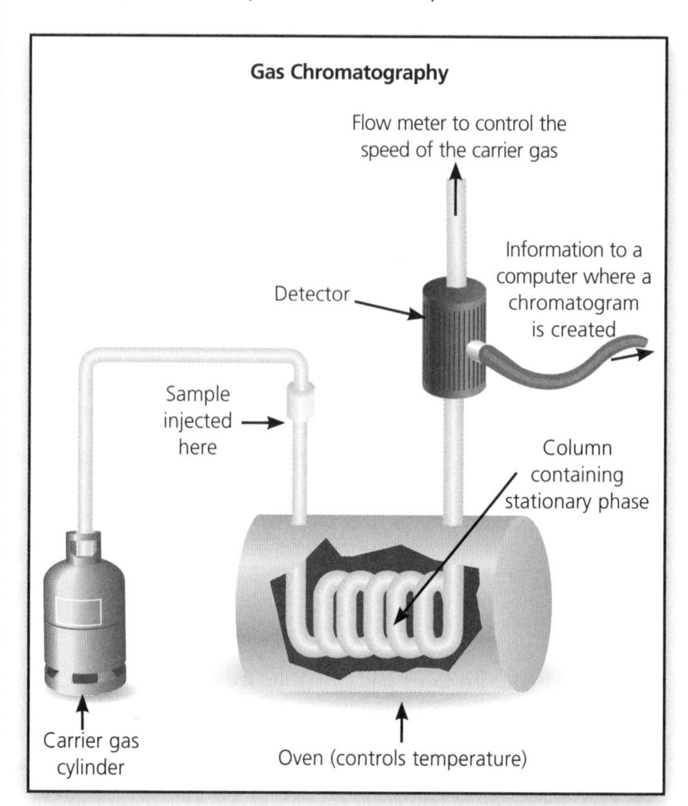

GC has a greater separating power than TLC or paper chromatography and can separate complex mixtures. It can produce quantitative data from very small samples of liquids, gases and volatile solids.

The size of each peak in the chromatogram produced by GC shows the relative amount of each chemical in the sample. For example, the chromatogram below shows six different compounds present in a sample.

It can be seen that Compound A is present in the largest amount and Compound D in the smallest amount.

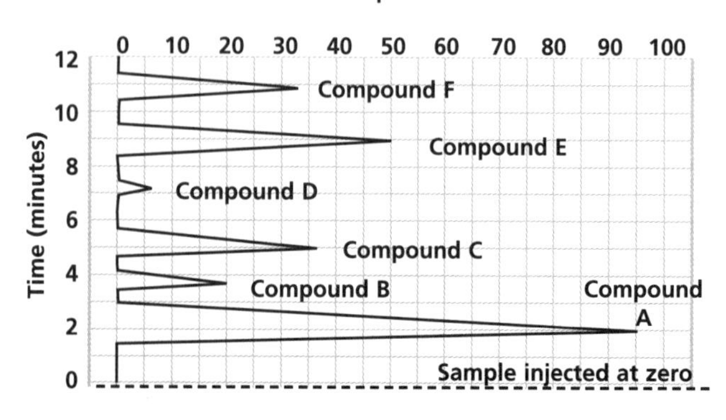

GC can separate the components in a mixture because of their different solubilities in the stationary or mobile phases.

Practical uses of GC include:

- detecting banned substances in blood and urine samples (e.g. random sampling of athletes)
- analysing the exact characteristics of oil or pesticide spills and matching them to samples from suspected sources, to identify where the pollution has come from.

The time taken for each substance to pass through the chromatographic system depends on its solubility. This is called the **retention time**.

In GC, the retention time is the time taken from the sample being injected into the system to when the substance is detected.

Tables of relative retention times show the retention times of different chemicals relative to the retention time of a specific compound.

Example

Methanol has a retention time of 2.24 minutes. Using the data below, which compound could be methanol?

Compound	Retention Time (minutes)
A	2.08
B	2.24
C	3.01
D	2.57

Compound B could be methanol as it has the same retention time.

Quantitative Analysis

Quantitative analysis determines the amount of a chemical in a sample. The flow chart opposite shows the main stages of any quantitative analysis.

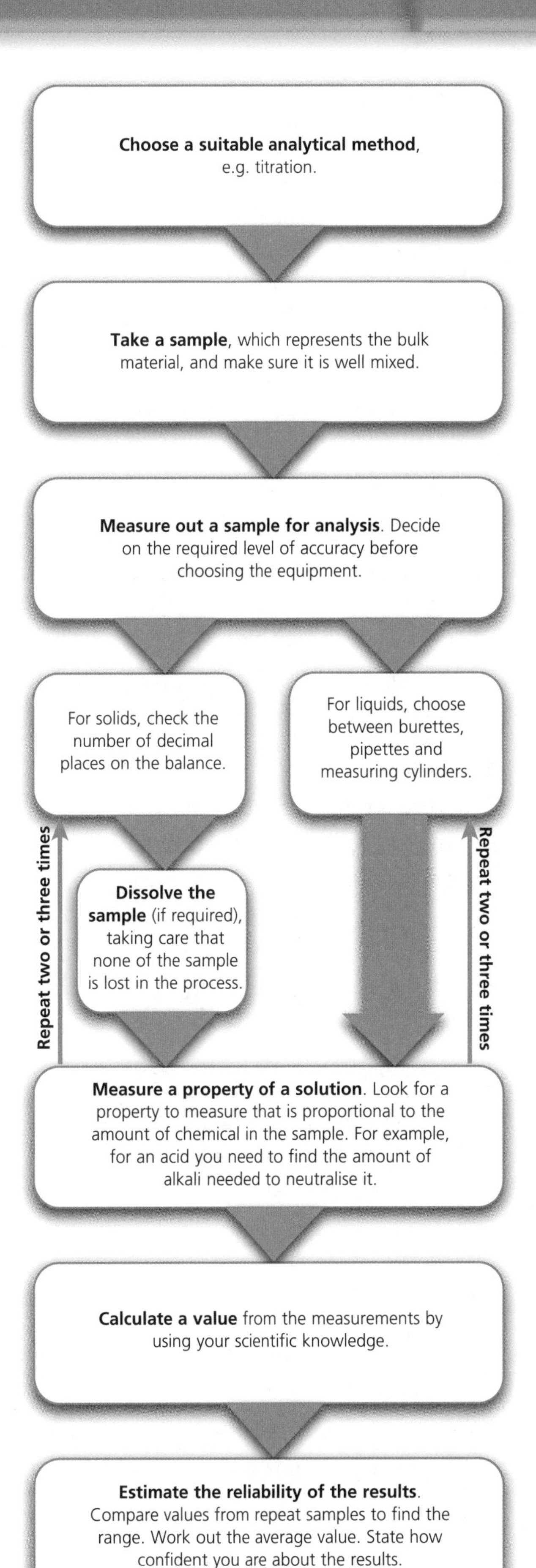

Calculating Concentration and Mass

Many methods of quantitative analysis use solutions. The **concentration** of a solution is the quantity of solid dissolved in the liquid. The concentration of a solution is measured in g/dm^3.

HT The formula below is used to calculate concentration:

$$\text{Concentration } (g/dm^3) = \frac{\text{Mass (g)}}{\text{Volume } (dm^3)}$$

Example 1
3.6g copper sulfate is dissolved in $80cm^3$ water. What is the concentration of the solution?

$$\text{Concentration} = \frac{\text{Mass}}{\text{Volume}}$$

Divide by 1000 to convert cm^3 to dm^3

$$= \frac{3.6g}{\left(\frac{80cm^3}{1000}\right)} = \mathbf{45g/dm^3}$$

Example 2
Calculate the concentration of the solution when 105g sodium chloride is dissolved in 3 litres of water.

$$\text{Concentration} = \frac{\text{Mass}}{\text{Volume}}$$

1 litre is the same as $1dm^3$

$$= \frac{105g}{3dm^3} = \mathbf{35g/dm^3}$$

Example 3
$100cm^3$ copper sulfate is prepared at a concentration of $52g/dm^3$. Calculate the mass of copper sulfate used.

Rearrange the formula:

Mass = Concentration × Volume

$$\text{Mass} = 52g/dm^3 \times \frac{100}{1000} dm^3 = \mathbf{5.2g}$$

Example 4
Calculate the mass of solute if the concentration of a solution is $42g/dm^3$ and the volume is $2dm^3$.

Mass = Concentration × Volume

$$\text{Mass} = 42g/dm^3 \times 2dm^3 = \mathbf{84g}$$

Standard Solutions

The concentrations of **standard solutions** are known accurately. Therefore, these solutions can be used to measure the concentration of other solutions. A standard procedure is used to make up the solution.

For example, the following method is used to make up a standard solution of copper sulfate:

1. Weigh out accurately a small amount of copper sulfate in a beaker (about 5g is suitable).
2. Transfer the solid copper sulfate into a volumetric flask using a short-stem funnel. Wash the funnel and beaker with distilled water. Pour the washings into the volumetric flask (this will ensure that all of the solid has been transferred).
3. Add distilled water to the flask until it is about three-quarters full. Place the stopper in the top and gently shake until all the solid is dissolved.
4. Place the flask on a level surface and fill it up with water until the level of solution reaches the $100cm^3$ mark.
5. Put a stopper in the flask and gently shake to ensure that the concentration is the same throughout the mixture.

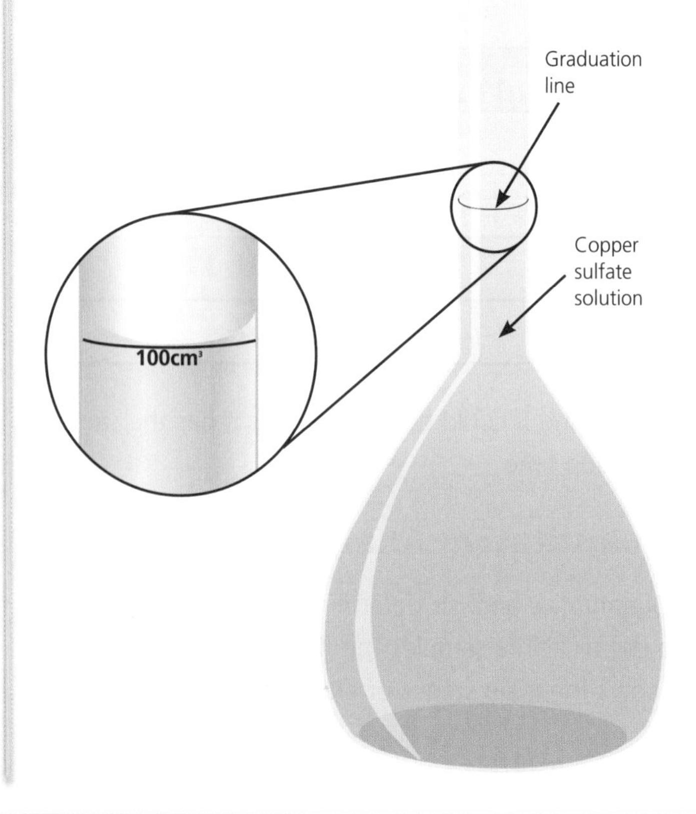

Quantitative Analysis by Titration

Acid–alkali titration is a very important method of quantitative analysis. You must be able to apply the procedure outlined below to different situations:

1. Fill a burette with the alkali (the concentration of the alkali must be known) and take an initial reading of the volume.
2. Accurately weigh out a 4g sample of solid acid and dissolve it in 100cm^3 of distilled water.
3. Use a pipette to measure 25cm^3 of the aqueous acid and put it into a conical flask. Add a few drops of an indicator (e.g. phenolphthalein) to the conical flask. (The indicator will show its acidic colour.) Place the flask on a white tile.
4. Add the alkali from the burette to the acid in the flask drop by drop. Swirl the flask to ensure it mixes well. Near the end of the reaction, the indicator will start to turn the alkali colour (e.g. pink for phenolphthalein). Keep swirling and adding the alkali until the indicator goes permanently pink on the addition of one drop of alkali, showing that the acid has been neutralised.
5. Record the volume of the alkali added by subtracting the initial burette reading from the final burette reading.
6. Repeat the whole procedure until you get two results that are the same or within 0.05cm^3 of one another. Alternatively, repeat the procedure three times and take the average.

As well as an indicator, a pH probe can also be used to measure the change in pH. The end point of the reaction can be determined from a pH / volume graph.

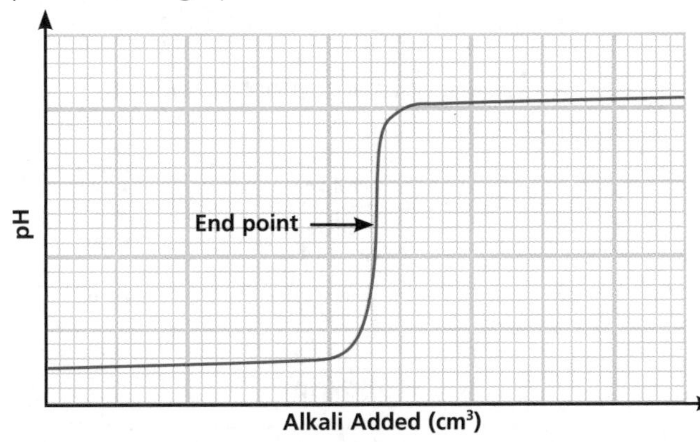

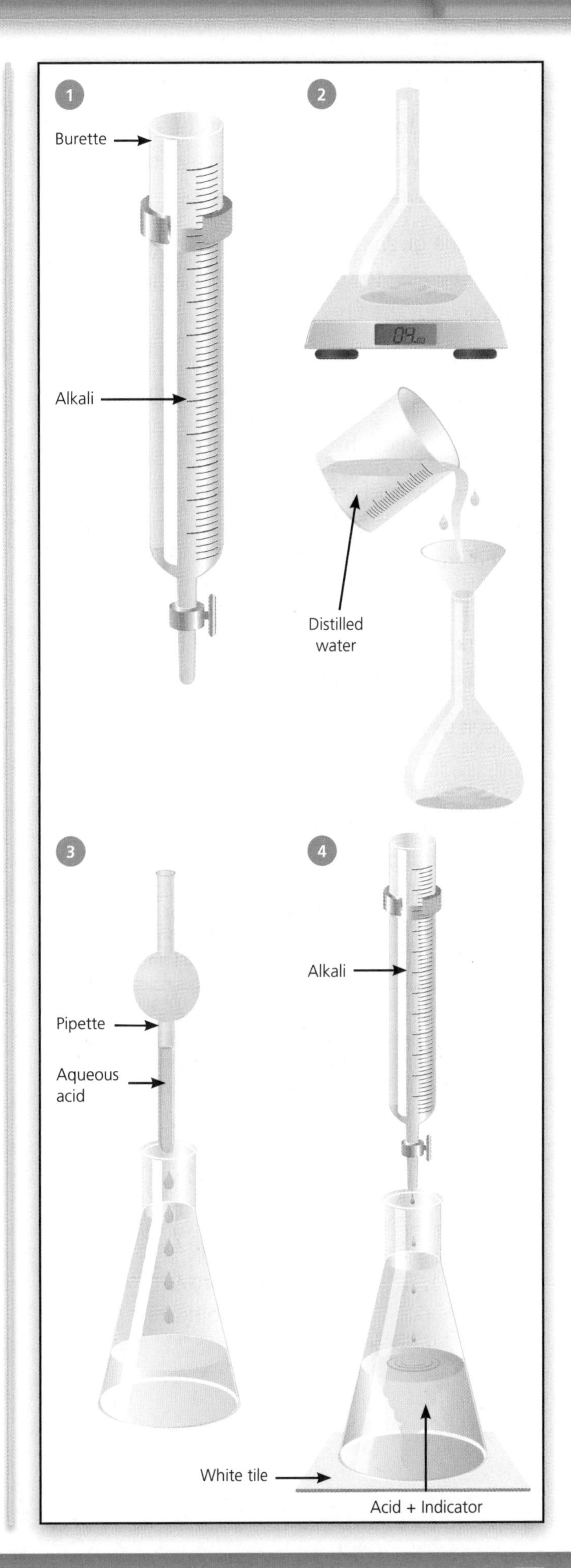

Interpreting Titration Results

When asked to interpret the results of a titration experiment, you will be given all the information required to carry out the calculation.

You will be given the titration formula and you need to be able to substitute the correct numbers and work out the final answer.

Example

Concentration of sodium hydroxide = 30g/dm^3

Volume of sodium hydroxide = 25cm^3

Volume of hydrochloric acid added:
Experiment 1: 10.05cm^3
Experiment 2: 10.00cm^3
Experiment 3: 9.95cm^3

(a) Work out the average volume of hydrochloric acid used in the three experiments.

$$\text{Average volume} = \frac{10.05\text{cm}^3 + 10.00\text{cm}^3 + 9.95\text{cm}^3}{3} = 10.0\text{cm}^3$$

(b) Using the formula given below, calculate the concentration of hydrochloric acid.

$$\text{Concentration of acid} = \frac{\text{Volume of NaOH (dm}^3) \times \text{Conc. of NaOH (g/dm}^3)}{\text{Volume of acid (dm}^3)} \times 0.9125$$

You must work in dm^3 when doing concentration calculations. To convert cm^3 to dm^3, divide by 1000.

$$\text{Concentration of hydrochloric acid} = \frac{\left(\frac{25}{1000}\right) \times 30 \times 0.9125}{\left(\frac{10}{1000}\right)}$$

= **68g/dm^3** (answer rounded to nearest g/dm^3)

HT You must be able to interpret the results of a titration using a balanced equation and the relative formula masses (see page 46). Use the steps in the worked example below.

Example

A titration is carried out and 35cm^3 of sulfuric acid of concentration 60g/dm^3 neutralises 25cm^3 of sodium hydroxide. Calculate the concentration of sodium hydroxide.

1. Work out the relative formula masses of the acid and alkali.

   $H_2SO_4 = (2 \times 1) + 32 + (4 \times 16) = 98$

   $NaOH = 23 + 16 + 1 = 40$

2. Write down the equation.

   $H_2SO_{4(aq)} + 2NaOH_{(aq)} = Na_2SO_{4(aq)} + 2H_2O_{(l)}$

   H_2SO_4 ↓ **98**; $2NaOH$ ↓ **2 × 40**

   This means that 98g of sulfuric acid reacts with 80g of sodium hydroxide.

3. Work out the mass of sulfuric acid used in the titration.

   Mass = Concentration × Volume

   $= 60\text{g/dm}^3 \times \left(\frac{35\text{cm}^3}{1000}\right)$

   $= 2.1\text{g}$

4. Work out the mass of sodium hydroxide used in the reaction.

   If 98g of sulfuric acid reacts with 80g of sodium hydroxide, then 2.1g reacts with $\frac{2.1}{98} \times 80\text{g} = 1.7\text{g}$ sodium hydroxide.

5. Work out the concentration of sodium hydroxide.

   $$\text{Concentration} = \frac{\text{Mass}}{\text{Volume}} = \frac{1.7\text{g}}{\left(\frac{25\text{cm}^3}{1000}\right)}$$

   = **68g/dm^3**

Evaluating Experimental Results

The validity of an experiment can depend on the **accuracy** of the results. Accuracy describes how close a result is to the true value or 'actual' value.

Inaccurate results can be the result of errors of measurements or mistakes. Mistakes are errors that are introduced when the person undertaking the experiment does something incorrectly, for example:

- misreading a scale
- forgetting to fill up a burette to the correct level
- taking a thermometer out of the solution to read the scale.

There are two general sources of measured uncertainty: **systematic** errors and **random** errors.

Precision is a measure of the spread of the measured values. A big spread leads to a greater uncertainty.

The degree of uncertainty is often assessed by working out the average results and stating the range.

Example

In an experiment the following repeat measurements of a concentration were taken: 72.0g/dm^3, 72.4g/dm^3, 71.9g/dm^3, 72.1g/dm^3, 71.8g/dm^3

Calculate the average result and degree of uncertainty.

The average result:

$$\frac{72.0 + 72.4 + 71.9 + 72.1 + 71.8}{5} = \mathbf{72.04g/dm^3}$$

The range is from 71.8g/dm^3 to 72.4g/dm^3.

This gives an overall uncertainty of 0.6g/dm^3.

$$\text{Percentage error} = \frac{0.6\text{g/dm}^3}{72.04\text{g/dm}^3} \times 100$$

(Uncertainty → 0.6g/dm^3; Average → 72.04g/dm^3)

= 0.83%

Certainty is 100 – 0.83 = **99.17%**

This result may be quoted as 99.17+/– 0.83% certain.

Systematic Errors

Systematic errors mean that repeat measurements are consistently too high or low. This could result from an incorrectly calibrated flask.

For example, a burette reading could be +/– 0.05cm^3. In the diagram below the burette reading is 0.05cm^3 out due to poor calibration. This means that when the meniscus is on the line, the actual volume is 25.05cm^3.

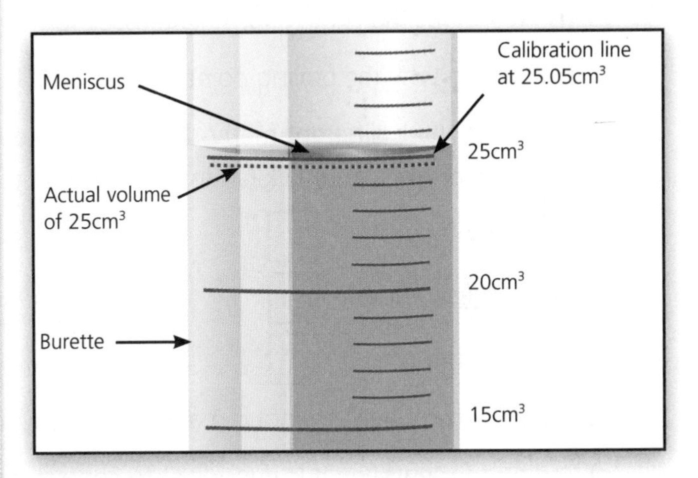

If the burette is used at a different temperature from the temperature it was calibrated at, then a systematic error might be introduced.

Random Errors

Random errors mean that repeat measurements give different values. For example, repeat measurements can introduce random errors because the meniscus is not on the calibration line.

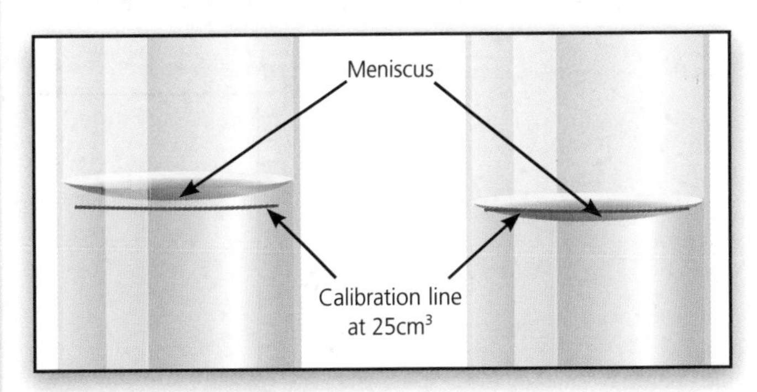

The end point of a titration can be determined by using a pH meter or light sensor. It can also be detected using the naked eye, but this method may introduce random errors.

Exam Practice Questions

1. Ethanol is a very important industrial chemical. It can be used as a solvent, a fuel or a component in alcoholic drinks.

   (a) The diagram shows the structural formula of ethanol.

```
    H   H
    |   |
H — C — C — O — H
    |   |
    H   H
```

   On the diagram, draw a circle around the functional group that is responsible for the characteristic properties of alcohols. **[1]**

   (b) Ethanol has a boiling point of 78.5°C, whereas ethane has a boiling point of -88.6°C.

   (i) Work out the relative formula masses for ethanol and ethane. **[2]**

   (ii) Explain why the boiling point of ethanol is higher than expected. **[2]**

   (c) There are three different methods used to produce ethanol. Which method is usually used to produce alcoholic drinks? Put a tick (✓) in the box next to the correct answer. **[1]**

   Biotechnology ☐

   Chemical synthesis ☐

   Fermentation ☐

   (d) Explain why, when ethanol is manufactured by fermentation, the concentration is limited. **[2]**

   (e) During fermentation it is important to control the temperature and pH of the reaction. Explain why. **[4]**

   (f) The synthetic method of producing ethanol is the least sustainable of the three production methods. It uses non-renewable feedstock, high temperatures, high pressures and a catalyst.

   Using your knowledge of the principles of green chemistry, explain what could be done to make this process more sustainable. **[4]**

2. A customs officer finds a bag of white powder when searching through a traveller's rucksack. The traveller tells him it is washing powder. The substance is sent to a forensic laboratory so that it can be analysed using chromatography.

   When the chromatogram was complete, the forensic scientist sprayed it so that the spots could be seen clearly.

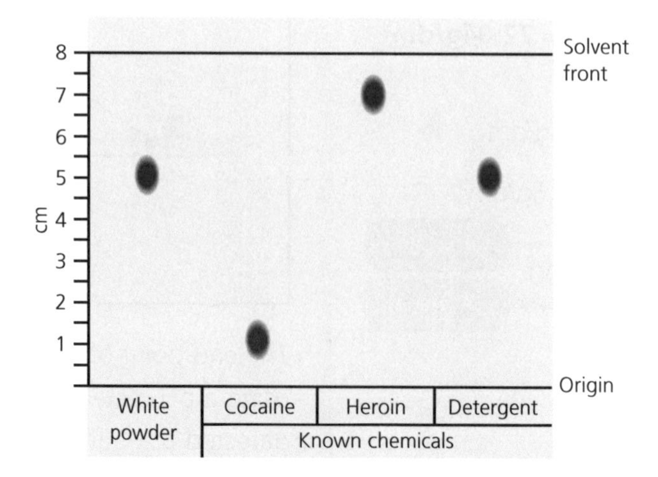

(a) What was the chromatogram sprayed with? Put a tick (✓) in the box next to the correct answer. **[1]**

Locating agent ☐ Reducing agent ☐

Oxidising agent ☐ Bleaching agent ☐

(b) Use the formula below to calculate the R_f value for each spot. **[3]**

$$R_f \text{ value} = \frac{\text{Distance travelled by substance}}{\text{Distance travelled by solvent}}$$

(c) Which substance was in the white powder? **[1]**

(d) Explain, in as much detail as you can, how paper chromatography works. Include the following terms in your answer. **[6]**

Mobile phase Stationary phase Dynamic equilibrium Molecules Solvent Attracted

The quality of written communication will be assessed in your answer to this question.

HT

3 When methane gas burns and oxygen is present, carbon dioxide, water, and heat and light energy are given out.

(a) Complete and balance the equation for the reaction. **[2]**

CH_4 + ... ⟶ CO_2 + ...

(b) Calculate the mass of CO_2 produced by burning one kilogram of CH_4. **[3]**

(Relative atomic masses: C – 12, H – 1, O – 16)

(c) In terms of the bonds between atoms and molecules, describe what happens during the chemical reaction when methane burns. Include the following words in your answer.

Activation energy Endothermic Exothermic **[4]**

(d) Which energy-level diagram best explains what happens during the burning of methane? Put a tick (✓) in the box under the correct diagram. **[1]**

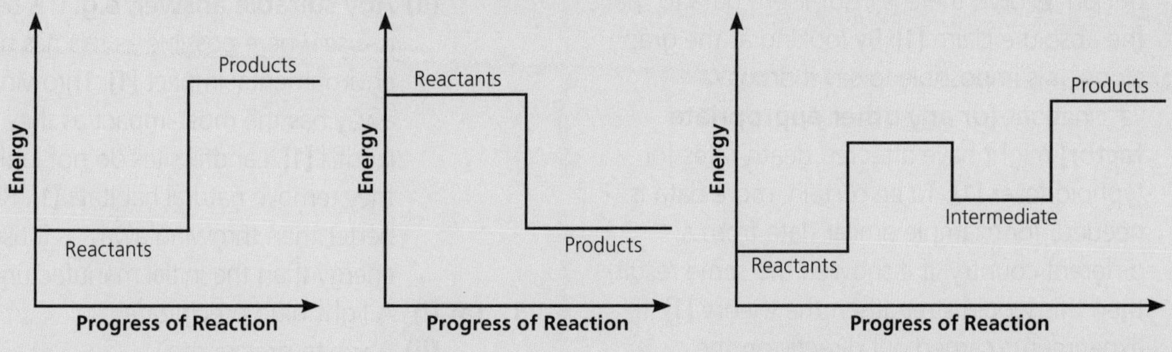

☐

☐

☐

Answers

Unit A171 (Pages 27–28)

1. (a) | B | A | D | E | C |
   |---|---|---|---|---|

   [1 mark if one correct, 2 marks if two correct, 3 marks if all correct.]

   (b) C **should be ticked.**

   (c) **This is a model answer which would score full marks:** Carbon dioxide levels are increasing. This is mainly due to the burning of fossil fuels. The fossil fuels, coal and natural gas, are used in power stations to generate electricity. In modern life, we use many electrical appliances every day for work and leisure, e.g. computers and games consoles. Most motor vehicles are powered by a combustion engine that is fuelled by petrol or diesel, which are produced from fossil fuels. Deforestation is another man-made problem that is leading to increased levels of carbon dioxide. Forests play an important role in the natural removal of carbon dioxide during photosynthesis. As forests are cut down, this natural carbon dioxide sink is being removed.

2. (a) 1908

   (b) The numbers generally decrease from about 24 per 100 000 population in 1908 to zero in 1945 **[1]**. The most rapid decline in deaths was seen in the first 10 years after chlorine was introduced, when the number of deaths decreased by over half **[1]**.

   (c) The chlorine killed the microorganism present in the water that caused typhoid fever.

   (d) The data in the graph does show a correlation between chlorinated water and typhoid fever, but on its own there is insufficient data to make the absolute claim **[1]**. By looking at the graph alone, it is impossible to say if drugs / vaccinations **[or any other appropriate factor]** might have affected death rates for typhoid fever **[1]**. To be certain, more data is needed, for example similar data from a different country. If it showed the same result, then this would strengthen the theory **[1]**. If experiments carried out directly on the microorganism responsible for typhoid fever showed positive results, this would add further evidence of a correlation **[1]**.

3. Hydrogen gas; Chlorine gas; **and** Sodium hydroxide **should be ticked.**

4. (a) The biodegradable bags are considerably weaker than the other two.

   (b) Find the range of the values, disregarding any outliers **[1]**, and calculate the mean (average) **[1]**.

   (c) 26.70N

   (d) (26.50 + 26.53 + 26.49) ÷ 3 = 26.51N **[1 mark for correct working but wrong answer.]**

   (e) (23.48 + 23.45 + 23.49 + 23.46) ÷ 4 = 23.47N **[1 mark for correct working but wrong answer.]**

   (f) **This is a model answer which would score full marks:** The manager should choose the polythene bag because it is stronger than the UV-degradable bag and will allow people to carry heavier shopping. Biodegradable bags are even weaker than UV-degradable bags, so would be less useful for carrying heavy shopping. Some members of the public think UV-degradable bags are the strongest, so the manager will have to explain that this is not what the data suggests. Polythene is not biodegradable, but it can be re-used and eventually recycled. Therefore, it may never end up in landfill. **[The UV-degradable bag could also be given as an acceptable answer, so long as it is well written and supported by valid arguments.]**

Unit A172 (Pages 63–64)

1. (a) (i) H_2O (ii) CH_4

   (b) C + 2 O = 12 + (2 × 16) = 44 **[1 mark for correct working but wrong answer.]**

2. (a) Reduction **should be ringed.**

   (b) (i) Re-use; Recycle; Throw away / landfill

   (ii) **Any suitable answer, e.g.** It is best to re-use where possible as this has no environmental impact **[1]**. Throwing things away has the most impact as they will go to landfill **[1]**. Landfill sites do not look good and they remove natural habitats **[1]**. Recycling is better than throwing away as it uses a lot less energy than the initial manufacturing **[1]**.

3. (a) (i) A light blue precipitate

   (ii) A white precipitate

   (b) Iron(III) iodide **[1]**. A red-brown precipitate forming with sodium hydroxide is the positive test for iron(III) **[1]**. A yellow precipitate forming with acidified silver nitrate is the positive test for iodide ions **[1]**.

4. **(a)** Sodium reacts with cold water **and** Sodium tarnishes in air to make sodium oxide **should be ticked.**
   (b) (i) 11 protons; 12 neutrons
   (ii) The diagram should have two crosses on the first shell, eight on the second shell and one on the third shell.
   (iii) 2.8.1
   (c) (i) Any two from: Work in a fume cupboard; Wear safety goggles; Wear protective clothing, e.g. gloves, coat; Use a safety screen.
   (ii) $2Na + Cl_2 \longrightarrow 2NaCl$ **[1 mark for correct symbols; 1 mark for 2Na.]**
   (iii) In this order: (s); (g); (s)
   (d) This is a model answer which would score full marks: The electron in the outer shell in the sodium atom will go into the outer shell of the chlorine atom. This will leave the sodium atom with a positive charge and the chlorine atom with a negative charge. The sodium and chloride ions are attracted together to form a crystal lattice of sodium chloride.
   (e) The reactivity of alkali metals increases as you go down the group **[1]**. This is because the outermost electron is further away from the influence of the nucleus and so an electron is lost more easily **[1]**. Potassium is below sodium in Group 1, so the reaction between potassium and chlorine is faster **[1]**.

Unit A173 (Pages 88–89)

1. **(a)** The OH group **should be circled**.
   (b) (i) Ethanol = (2 × 12) + 16 + 6 = 46; Ethane = (2 × 12) + 6 = 30
   (ii) The attractive forces between ethanol molecules are greater than those between ethane molecules **[1]** because the –OH group in ethanol makes the molecule polar, like water, resulting in stronger attractive forces between the molecules **[1]**.
   (c) Fermentation **should be ticked.**
   (d) The amount of sugar and enzymes present are limiting factors **[1]**. The reaction will stop if all the sugar is used up or if the concentration of alcohol reaches about 15%, which damages the enzymes and stops them from working. **[1]**.
   (e) Any suitable answer, e.g. The enzymes become denatured and are no longer able to work **[1]** if the temperature rises too high **[1]**. The pH must also be carefully controlled because the enzyme can become denatured by a change in pH **[1]** due to the attractions of excess H^+ ions or OH^- ions for the enzyme **[1]**.
   (f) Any suitable answer, e.g. A renewable feedstock could be sought **[1]**; Use a renewable source of energy to power the plant **[1]**; Find a more efficient catalyst that would allow the reaction to take place at lower temperatures **[1]** and lower pressures **[1]**.
2. **(a)** Locating agent **should be ticked.**
   (b) White powder: 0.63
   Cocaine: 0.13
   Heroin: 0.88
   Detergent: 0.63
   [3 marks if all correct; 2 marks if one wrong.]
   (c) Detergent
   (d) This is a model answer which would score full marks: A chromatogram is formed when chemicals come out of solution and bind to the paper. For each component of the sample, a dynamic equilibrium is set up between the stationary phase (paper) and the mobile phase (solvent). Different molecules in the sample travel different distances according to how strongly they are attracted to the molecules in the stationary phase, in relation to their attraction to the solvent phase.
3. **(a)** $2O_2$; $2H_2O$
   (b) RFM of CH_4 = 16; RFM of CO_2 = 44 **[1]**; 16g of CH_4 produce 44g of CO_2 **[1]**; 1kg of CH_4 produces 2.75kg of CO_2 **[1]**.
   (c) Any suitable answer, e.g. To start the reaction between methane and oxygen, the activation energy must be reached **[1]**. This is the energy that is needed to break the O=O bonds in oxygen and the C–H bonds in methane **[1]**. This part of the reaction is endothermic because the energy needed to break the bonds is taken in from the surroundings **[1]**. The atoms then rearrange, and new bonds are formed between different atoms. During this part of the reaction, energy is given out, so it is exothermic **[1]**.
   (d) The second diagram should be ticked.

Glossary

Acid – an aqueous compound with a pH value less than 7.

Activation energy – the minimum amount of energy required to cause a reaction.

Alkali – a substance that has a pH value higher than 7.

Aqueous solution – a solution made when a solute dissolves in water.

Atmosphere – the layer of gases surrounding the Earth.

Atom – the smallest part of an element that can enter into a chemical reaction.

Atom economy – the proportion of reactants that are converted into useful products rather than waste products.

Biodegradable – a word that describes a material that can be broken down by bacteria.

Brown atom – a reacting atom that ends up in a waste product.

Bulk chemicals – chemicals made on a large scale.

Catalyst – a substance that is used to speed up a chemical reaction without being chemically altered itself.

Chemical synthesis – the process by which many useful products are made.

Chromatography – a technique used to separate different compounds in a mixture according to how well they dissolve in a particular solvent.

Combustion – a chemical reaction that occurs when fuels burn, releasing heat.

Compound – a substance in which the atoms of two or more elements are chemically joined, either by ionic or covalent bonds.

Covalent bond – the force of attraction between two atoms sharing electrons.

Cross-links – strong links between polymer chains.

Crystallisation – the formation of solid crystals from a solution.

Decomposer – Organisms such as bacteria that break down dead or dying organic matter.

Denatured – the state of an enzyme that has been destroyed by heat or pH and can no longer work.

Distillate – the product of distillation.

Distillation – the process of separating a liquid from a mixture by boiling the mixture to evaporate the liquid and then condensing the vapours.

Electrolysis – the process by which an electric current causes a solution, containing ions, to undergo chemical decomposition.

Electron – a negatively charged particle that orbits the nucleus.

Element – a substance that consists of one type of atom.

Endothermic reaction – a chemical reaction that takes in energy (heat) from its surroundings so that the products have more energy than the reactants.

Enzyme – a protein molecule and biological catalyst found in living organisms that helps chemical reactions to take place (usually by increasing the rate of reaction).

Equilibrium – when a reversible reaction appears to have 'stopped', as the concentration of products and reactants are not changing.

Evaporation – the process in which a liquid changes into a gas by heating.

Exothermic reaction – a chemical reaction that gives out energy (heat) to its surroundings so that the products have less energy than the reactants.

Filtration – a method for separating solids from liquids by passing a mixture through a porous material.

Fine chemicals – chemicals manufactured on a small scale.

Fossil fuel – fuels formed in the ground, over millions of years, from the remains of dead plants or animals.

Fuel – a substance that releases energy when burned in the presence of oxygen.

Global warming – the increase in the average temperature on Earth due to a rise in the level of greenhouse gases in the atmosphere.

Green atom – a reacting atom that ends up in a useful product.

Green chemistry – the production of chemicals based on principles that can lead to a more sustainable process.

Greenhouse gas – gases in the Earth's atmosphere that absorb radiation and stop it from leaving the Earth's atmosphere.

Group – a vertical column of elements in the periodic table.

Halogen – an element found in Group 7 of the periodic table.

Homologous series – a family of organic compounds with similar chemical properties.

Hydrocarbon – a compound made of carbon and hydrogen atoms only.

Hydrosphere – contains all the water on Earth, including rivers, oceans, lakes, etc.

Insoluble – a property that means a substance cannot dissolve in a solvent.

Ion – a positively or negatively charged particle formed when an atom, or group of atoms, loses or gains electrons.

Life cycle assessment (LCA) – an analysis of a product from manufacture to disposal.

Lithosphere – the rigid outer layer of the Earth made up of the crust and the part of the mantle just below it.

Mobile phase – the substance that moves in a definite direction during chromatography. It consists of the sample being separated and the solvent.

Monomers – small molecules that join together to form polymers.

Nanoparticle – a particle that is less than 100nm in size.

Nanoscale – things relating to or occurring on a scale of nanometres.

Nanoscience – the science of structures that are 1–100 nanometres in size.

Nanotechnology – a branch of technology dealing with the manufacture of objects with dimensions of less than 100nm and the manipulation of individual molecules and atoms.

Neutralisation – the reaction between an acid and a base which forms products that are pH neutral.

Neutron – a particle found in the nucleus of an atom that has no electric charge.

Nitrogen fixation – the conversion of nitrogen gas into nitrogen compounds such as nitrates and ammonia.

Non-renewable resources – resources (especially energy sources) that cannot be replaced in a lifetime.

Outlier – a measured result that appears to be very different from the value you would expect or from other measured results. Therefore you strongly suggest that it is wrong.

Oxidation – a chemical reaction that occurs when oxygen joins with an element or compound.

Period – a horizontal row of elements in the periodic table.

Photosynthesis – the chemical process that takes place in green plants where water combines with carbon dioxide to produce glucose using light energy.

Glossary

Pollutant – a chemical that can harm the environment and health.

Polymer – a long-chain hydrocarbon molecule built up from small units called monomers.

Precipitate – an insoluble solid formed during a reaction involving solutions.

Precipitation – the process of forming a precipitate by mixing solutions.

Proton – a positively charged particle found in the nucleus of an atom.

Recycling – to re-use materials that would otherwise be considered as waste.

Reduction – a chemical reaction that occurs when oxygen is removed.

Relative atomic mass (RAM or A_r) – the average mass of an atom of an element compared with a twelfth of the mass of a carbon atom.

Relative formula mass (RFM or M_r) – the sum of the atomic masses of all the atoms in a molecule.

Renewable resources – resources that can be replaced as quickly as they are used up.

Residue – the substance that remains after a chemical reaction or a process (e.g. filtration).

R_f value – the movement of a substance relative to the movement of the solvent front.

Salt – the product of a chemical reaction between a base and an acid.

Saturated molecule – organic molecules where the carbon–carbon bonds are single.

Soluble – a property that means a substance can dissolve in a solvent.

Solution – the mixture formed when a solute dissolves in a solvent.

Solvent – a liquid that can dissolve another substance to produce a solution.

Standard reference material – a material that has known properties and can be used as a control.

Stationary phase – the substance fixed in place in chromatography, e.g. in paper chromatography it is the paper.

Sustainable – capable of being continued with minimal long-term effect on the environment; resources that can be replaced or maintained.

Titration – an accurate technique that can be used to find the volume of liquid needed to neutralise an acid.

Universal indicator – a mixture of pH indicators, which produces a range of colours according to pH and can, therefore, be used to measure the pH of a solution.

Unsaturated molecule – organic molecules containing at least one carbon–carbon double bond.

Yield – the amount of product obtained, e.g. from a crop or a chemical reaction.

HT **Reflux** – a process of continuous heating without the loss of volatile substances.

Tap funnel – glass funnel with a tap, used for separating immiscible liquids.

Wet scrubbing – a method of cleaning exhaust air or flue gases.

Key

relative atomic mass
atomic symbol
name
atomic (proton) number

1	2											3	4	5	6	7	0
						1 **H** hydrogen 1											4 **He** helium 2
7 **Li** lithium 3	9 **Be** beryllium 4											11 **B** boron 5	12 **C** carbon 6	14 **N** nitrogen 7	16 **O** oxygen 8	19 **F** fluorine 9	20 **Ne** neon 10
23 **Na** sodium 11	24 **Mg** magnesium 12											27 **Al** aluminium 13	28 **Si** silicon 14	31 **P** phosphorus 15	32 **S** sulfur 16	35.5 **Cl** chlorine 17	40 **Ar** argon 18
39 **K** potassium 19	40 **Ca** calcium 20	45 **Sc** scandium 21	48 **Ti** titanium 22	51 **V** vanadium 23	52 **Cr** chromium 24	55 **Mn** manganese 25	56 **Fe** iron 26	59 **Co** cobalt 27	59 **Ni** nickel 28	63.5 **Cu** copper 29	65 **Zn** zinc 30	70 **Ga** gallium 31	73 **Ge** germanium 32	75 **As** arsenic 33	79 **Se** selenium 34	80 **Br** bromine 35	84 **Kr** krypton 36
85 **Rb** rubidium 37	88 **Sr** strontium 38	89 **Y** yttrium 39	91 **Zr** zirconium 40	93 **Nb** niobium 41	96 **Mo** molybdenum 42	[98] **Tc** technetium 43	101 **Ru** ruthenium 44	103 **Rh** rhodium 45	106 **Pd** palladium 46	108 **Ag** silver 47	112 **Cd** cadmium 48	115 **In** indium 49	119 **Sn** tin 50	122 **Sb** antimony 51	128 **Te** tellurium 52	127 **I** iodine 53	131 **Xe** xenon 54
133 **Cs** caesium 55	137 **Ba** barium 56	139 **La*** lanthanum 57	178 **Hf** hafnium 72	181 **Ta** tantalum 73	184 **W** tungsten 74	186 **Re** rhenium 75	190 **Os** osmium 76	192 **Ir** iridium 77	195 **Pt** platinum 78	197 **Au** gold 79	201 **Hg** mercury 80	204 **Tl** thallium 81	207 **Pb** lead 82	209 **Bi** bismuth 83	[209] **Po** polonium 84	[210] **At** astatine 85	[222] **Rn** radon 86
[223] **Fr** francium 87	[226] **Ra** radium 88	[227] **Ac*** actinium 89	[261] **Rf** rutherfordium 104	[262] **Db** dubnium 105	[266] **Sg** seaborgium 106	[264] **Bh** bohrium 107	[277] **Hs** hassium 108	[268] **Mt** meitnerium 109	[271] **Ds** darmstadtium 110	[272] **Rg** roentgenium 111	Elements with atomic numbers 112–116 have been reported but not fully authenticated						

The lanthanoids (atomic numbers 58–71) and the actinoids (atomic numbers (90–103) have been omitted.

The relative atomic masses of copper and chlorine have not been rounded to the nearest whole number.

Qualitative Analysis

Tests for Positively Charged Ions

Ion	Test	Observation
Calcium Ca^{2+}	Add dilute sodium hydroxide	A white precipitate forms; the precipitate does not dissolve in excess sodium hydroxide
Copper Cu^{2+}	Add dilute sodium hydroxide	A light blue precipitate forms; the precipitate does not dissolve in excess sodium hydroxide
Iron(II) Fe^{2+}	Add dilute sodium hydroxide	A green precipitate forms; the precipitate does not dissolve in excess sodium hydroxide
Iron(III) Fe^{3+}	Add dilute sodium hydroxide	A red-brown precipitate forms; the precipitate does not dissolve in excess sodium hydroxide
Zinc Zn^{2+}	Add dilute sodium hydroxide	A white precipitate forms; the precipitate dissolves in excess sodium hydroxide

Tests for Negatively Charged Ions

Ion	Test	Observation
Carbonate CO_3^{2-}	Add dilute acid	The solution effervesces; carbon dioxide gas is produced (the gas turns limewater from colourless to milky)
Chloride Cl^-	Add dilute nitric acid, then add silver nitrate	A white precipitate forms
Bromide Br^-	Add dilute nitric acid, then add silver nitrate	A cream precipitate forms
Iodide I^-	Add dilute nitric acid, then add silver nitrate	A yellow precipitate forms
Sulfate SO_4^{2-}	Add dilute nitric acid, then add barium chloride or barium nitrate	A white precipitate forms

Notes

Index

Index